T0298345

# Pesticides, Organic Contaminants, and Pathogens in Air

# Pesticides, Organic Contaminants, and Pathogens in Air
## Chemodynamics, Health Effects, Sampling, and Analysis

James N. Seiber and Thomas M. Cahill

CRC Press
Taylor & Francis Group
Boca Raton London New York

CRC Press is an imprint of the
Taylor & Francis Group, an **informa** business

First edition published 2022
by CRC Press
6000 Broken Sound Parkway NW, Suite 300, Boca Raton, FL 33487-2742

and by CRC Press
2 Park Square, Milton Park, Abingdon, Oxon, OX14 4RN

CRC Press is an imprint of Informa UK Limited

ISBN: 978-0-367-48567-2 (hbk)
ISBN: 978-1-032-10894-0 (pbk)
ISBN: 978-1-003-21760-2 (ebk)

DOI: 10.1201/9781003217602

Typeset in Palatino
by codeMantra

*This book is dedicated to our colleagues who died after
the book project was launched.*

**Dwight Glotfelty, PhD,** was an expert in measurement of pesticides in air, rain, and fog, as well as in conducting flux studies in soil. A native of western Maryland, Glotfelty earned a PhD in environmental chemistry at the University of Maryland. He was a scientist in the Beltsville Agricultural Research Service, USDA, working with Dr. Alan Taylor, specializing in environmental analytical chemistry of pesticides in water, air, soil, and biota. He was coauthor of "Pesticides in Fog," published in *Nature* in 1987, and coauthor of a number of papers on pesticide flux to air during volatilization events.

**Michael Majewski, PhD,** was an expert in measuring and interpreting flux of contaminants in the air. He earned his PhD in agricultural and environmental chemistry at the University of California, Davis (UC, Davis), and held postdoctoral research positions at USDA-Agricultural Research Service, Riverside, California, and the Natural Resources Conservation Service in Sacramento, California.

**William Barry Wilson, PhD,** was a professor of avian sciences at UC, Davis. He was active in the pharmacology and toxicology graduate program at UC, Davis, and an expert on cholinesterase inhibition and inhibition of neurotoxic esterase. He was a spokesman for students during student antiwar demonstrations in the 1960s.

**Fumio Matsumura, PhD,** was known internationally as a "grand master of insect toxicology." His pioneering work shaped the fields of pesticide and environmental toxicology. Some of his major contributions were the mechanisms of action of tetrachlorodibenzo-$p$-dioxin (TCDD), the endocrine disruptive activities of DDT analogues, and many other topics. He was a distinguished professor of environmental toxicology and entomology at UC, Davis. Before that, he was a professor at the University of Wisconsin.

**Thomas A. Cahill** was a professor of physics, at UC, Davis. He is well known for his air quality research in such diverse situations as Mono Lake dust, wildfires, winter haze layers in the Grand Canyon, volcanic eruptions, oil rig fires, and emissions associated with the World Trade Center terrorist-instigated collapse and subsequent burning. He was one of the founding members of the IMPROVE air quality network, which was one of the first air quality networks; It is still operating today. Thomas also served as the Director of the UC Ecology Center. He was consulted on environmental safety and health issues worldwide.

**Lincoln Brower, PhD,** was an expert in chemical ecology and monarch butterflies. He studied monarch butterfly migration and sequestration of cardiac glycosides from milkweeds for defense against predators. In the photo, Lincoln Brower and Carolyn Nelson, research associate at UC, Davis, are in the Sierra foothills, bagging milkweed plants to study the uptake of cardiac glycosides in adult monarchs reared as larvae on these plants. He earned a BA in biology at Princeton University and a PhD at Yale University. He was a professor at Amherst College (Massachusetts) and then at the University of Florida. He published widely in journals such as *Science, Nature,* and *Scientific American.*

# Contents

# Acknowledgments

First and foremost, we acknowledge the significant efforts of James Woodrow, who wrote large sections of the chapters on physicochemical properties (Chapter 3), modeling (Chapter 5), fog (Chapter 7), and fumigants (Chapter 8), and kept us alerted to the latest scientific advances.

The staff research associates, colleagues, and students who did the bulk of the research described herein made this book possible, including Linda Aston, Lynn Baker, Sogra Begum, Carolinda Benson, Janet Benson, Martin Birch, Fan Chen, Tom Cahill, Tom Cook, Donald Crosby, Michael David, John Dolan, Ibrahim El Nazr, Geraldo Ferreira, Julia Frey, John Finley, Ben Giang, Hassan G. Fouda, William O. Gauer, Kate Gibson, Dwight Glotfelty, Martina Green, Corey Griffith, Matt Hengel, Bruce Hermann, Dirk Holstege, Puttanna Honaganahalli, Carol Johnson, Ron Kelly, Sue Keydell, Loreen Kleinschmidt, Peter Landrum, Mark Lee, James Lenoir, Qing X. Li, Anne Lucas, Steve Madden, Michael Majewski, Jim Markle, Melanie Marty, Terry Mast, and Michael McChesney, for analysis and field work, Laura McConnell, Glenn Miller, Seyed Mirsatori, Kerry Nugent, Mariella Paz Obeso, Tom Parker, Martha Philips, Mike Purdy, Carolyn Roeske Nelson, Lisa Ross, John C. Sagebiel, Paul Sanders, Charlotte Schomburg, Sami Selim, Charles Shoemaker, Mark Stelljes, Paul Tuskes, Jeanette Van Emon, Teresa Wehner, Carol Weiskopf, Gordon Wiley, Barry Wilson, Jennifer Wing, Wray Winterlin, James Woodrow, Chad Wujcik, Jack Zabik, Mark Zambrowski, and many others. The contributions of these individuals are recognized in the references to all sections of the book.

JNS acknowledges the help of his wife, Rita, whose support was essential to the completion of this book and to his children, Charles, Christopher, and Kenneth, and his grandchildren.

Special acknowledgment goes to Margaret Baker and Jeffry Eichler, who did the bulk of the technical editing and Madie Matibag who provided support to JNS during his illness and recovery.

TMC acknowledges his father, Thomas A. Cahill, as both a supporting father and a scientific mentor as well as his mother Virginia Cahill and sister Catherine Cahill who were always supportive and provided excellent sounding boards for ideas. Additionally, TMC thanks his mentors during his career, namely Donald Mackay of Trent University, Daniel Anderson of UC, Davis, Judi Charles of UC, Davis, and especially James Seiber who co-authored this book and was his Dissertation advisor at the University of Nevada, Reno.

# Authors

James N. Seiber, Professor Emeritus, University of California, Davis (UC, Davis), is a native of Hannibal, Missouri. He earned a BA in chemistry at Bellarmine College (Louisville, Kentucky), a master's degree in chemistry at Arizona State University (Tempe) under Myron Caspar, and a PhD in natural products chemistry at Utah State University (Logan) under Professor Frank Stermitz. He was a research chemist at Dow Chemical in Midland, Michigan, and Pittsburg, California, working on process development for new pesticides.

Dr. Seiber joined UC, Davis, in 1969 as a professor of environmental toxicology and a chemist at the California Agricultural Experiment Station. He later served as associate dean for research in the College of Agricultural and Environmental Sciences. Dr. Seiber moved to the University of Nevada, Reno, where he served as the Director of the Center for Environmental Sciences and Engineering. In 1990, he joined the U.S. Department of Agriculture (USDA) Agricultural Research Service (ARS) in Albany, California. Dr. Seiber spent sabbatical leaves at the International Rice Research Institute in the Philippines, the U.S. Environmental Protection Agency (EPA) in Perrine, Florida, and the USDA–ARS in Beltsville, Maryland.

While JNS was attending the 230th National American Chemical Society (ACS) meeting in 2005, in Washington, D.C., Hurricane Katrina was hitting the Gulf coast. Meeting attendees watched on TV as the storm developed. JNS was standing and watching TV, as was Ed Knipling and others from the USDA–ARS. They were concerned about the potential for Katrina to harm the Southern Regional Research Center (SRRC) and its USDA employees there.

About 2 weeks later, Dr. Knipling asked JNS to serve as Acting Director at SRRC, since Pat Jordan, SRRC Director, had just retired. JNS would continue in his permanent job as Western Regional Research Center (WRRC) Director, but would need to relocate to New Orleans, LA (NOLA) for about 3 months. He agreed, with the caveat that every week or so, he could return to Davis to help his wife Rita and his UC, Davis research group (which would be in the capable hands of research associates, James Woodrow and Michael McChesney). On top of that, he had also just undertaken a position as Editor of *Journal of Agricultural and Food Chemistry*, but Loreen Kleinschmidt, his Editorial Assistant, could handle that.

He was met by Pat Jordan and the SRRC Location Administrative Officer (LAO), and they toured the SRRC and neighboring areas to assess the damage. The Katrina floodwaters from the "surge" had receded, but the damage was everywhere, particularly at the ground level and first floor. Mold was a major concern, and removal of sheetrock was underway.

"Why did I go? Sense of duty mainly. As a member of the Senior Executive Service of the U.S., I was obligated to take assignments when requested. But I also looked forward to the experience, the opportunity to help and to live in NOLA and Cajun country."

His main job was to support the Research Leaders (RLs), make sure the SRRC could be safely reoccupied and research could continue, and to put a "good face" on matters. SRRC had research units in mycotoxins, nutrition, Formosan termite control, cotton and wool improvement, and a new one being formed on biofuels (specifically, studying the conversion of sugarcane and cane wastes to bioethanol). While at SRRC, JNS traveled to Houma, LA and met with Ed Richard and visited bioconversion facilities in that area, including pilot plants for bioenergy production, as well as campuses at Louisiana State University (LSU) in Baton Rouge and University of New Orleans (UNO) in NOLA. JNS also communicated and met with key national program leaders like Joe Spence, Antoinette Betschart, Wilda Martinez, Frank Flora, Jim Lindsay, and others, and with Midsouth and Oxford, MS USDA scientists. "I was amazed at the optimism everywhere, except the lower Ninth Ward, which was near despair."

During the Spring 2006 ACS meeting held in NOLA, JNS worked with Pat Jordan and ACS staff to host a media tour of flooded areas so they could see the progress being made post-Katrina. "It was a busy time!"

After the collateral assignment at SRRC ended, JNS was asked to take another assignment as Acting Director of the Western Human Nutrition Research Center (WHNRC) at Davis. The WHNRC was relocating to Davis from the Presidio in San Francisco, so much of his time was spent working with Dean Charles Hess and UC, Davis department chairs on arranging space for WHNRC scientists in UC, Davis buildings. "Very challenging!" Eventually the new WHNRC building was completed and Dr. Lindsay Allen, WHNRC Director and ARS scientist, moved in. "It was a very successful transition."

After the collateral assignment with WHNRC, JNS was asked by Chancellor Larry Van der Hoef to serve as Acting Chair of UC, Davis department of Food Science and Technology (FST). FST was relocating from Cruess Hall to the new Robert Mondavi Institute (RMI) facilities on campus. "It was a very tough assignment, particularly arranging the move of the FST Pilot Food Processing Plant from Cruess Hall to RMI which I enjoyed, but was happy to relinquish when it ended after 3 years."

Then JNS resumed life as Journal Editor, emeritus Professor of Environmental Toxicology and FST, working with graduate students, and teaching, along with some service activities with California Environmental Protection Agency and ACS.

He received the Spencer Award and Sterling Hendricks Lectureship and Lifetime Achievement Award in Environmental Toxicology from the AGRO Chemical Division of the ACS, as well as the Lifetime Achievement Award from the Environmental Toxicology department at UC, Davis.

Dr. Seiber and his research associates have published 300 manuscripts, books, and book chapters. He and his wife, Rita, reside in Davis, California, where they operate a small vineyard and visit with three children and seven grandchildren and other relatives and friends. Dr. Seiber currently teaches agricultural and environmental chemistry and serves on committees with the USDA, the EPA, and CalEPA. JNS immersed himself in the writing of this book and fishing.

"And recovering from onset of Parkinson's disease—biggest challenge of my life!"

**Thomas M. Cahill** is an Associate Professor in the School of Mathematical and Natural Sciences at Arizona State University. He grew up in science as the son of a physics professor at UC, Davis, so there was little doubt he would end up in science. His childhood was full of plants, fossils, rocks, and family vacations to national parks. His love for the environment led him to earn a BS in wildlife and fisheries biology at UC, Davis. His master's degree was also at UC, Davis, measuring the impact of mercury in the birds of Clear Lake, California, which had a superfund site on the shores of the lake that had discharged mercury into the ecosystem. His PhD investigated the hydrochlorofluorocarbon degradation product trifluoroacetic acid in water bodies that might be impacted by this persistent pollutant. Post-doctoral experience involved both environmental modeling of chemical fate and transport with Dr. Don Mackay and ambient air sampling for acrolein, which is a common highly reactive chemical that frequently appears near the top of hazardous air assessments.

In addition to Dr. Cahill's aerosol and acrolein research, he is leading projects that are assessing the toxicity associated with abandoned mines in the Sonora Desert. Recently, he has undertaken natural products projects to identify antiviral compounds in plants. He works extensively with undergraduate students and primarily teaches analytical chemistry and toxicology.

Dr. Cahill has written 46 peer-reviewed articles, including the following:

Cahill, T.M. (2013). Annual cycle of size-resolved organic aerosol characterization in an urbanized desert environment. *Atmospheric Environment* 71, 226–233.

Cahill, T.M., Thomas, C.M., Schwarzbach, S.E., and Seiber, J.N. (2001). Accumulation of trifluoroacetate in seasonal wetlands. *Environmental Science and Technology* 35, 820–825.

Seaman, V.Y., Bennett, D.H., and Cahill, T.M. (2007). Origin, occurrence and source emission rate of acrolein in residential indoor air. *Environmental Science and Technology* 41, 6940–6946.

# List of Abbreviations

| | |
|---|---|
| **2,4,5-T** | 2,4,5-trichlorophenoxyacetic acid |
| **2,4-D** | 2,4-dichlorophenoxyacetic acid |
| **ACGIH** | American Conference of Governmental Industrial Hygienists |
| **AChE** | acetylcholine esterase |
| **ACS** | American Chemical Society |
| **ADI** | acceptable daily intake |
| **AG** | aerodynamic-gradient |
| **AMPA** | 2-amino-3-(3-hydroxy-5-methyl-isoxazol-4-yl) propanoic acid |
| **ATP** | adenosine triphosphate |
| **ATWA** | annual time weighted average |
| **BCF** | bioconcentration factor |
| **BETX** | benzene and ethyl benzene, xylene, toluene |
| **BHC** | benzene hexachloride |
| **CAL FIRE** | California Department of Forestry and Fire Protection |
| **CDPR** | California Department of Pesticide Regulation |
| **CARB** | California Air Resources Board |
| **CASCC** | Caltech active strand cloudwater collector |
| **CCD** | colony collapse disorder |
| **CFC** | chlorofluorocarbon |
| **CIAT** | 2-chloro-4-isopropylamino-6-amino-s-triazine |
| **CIMIS** | California Irrigation Management Information Systems |
| **COVID-19** | coronavirus disease of 2019 |
| **DBCP** | dibromochloropropane |
| **DDE** | dichlorodiphenyldichloroethylene |
| **DDT** | dichlorodiphenyltrichloroethane |
| **EDB** | ethylene dibromide |
| **EF** | enrichment factor |
| **EPA** | Environmental Protection Agency |
| **EPTC** | S -ethyl dipropylthiocarbamate |
| **EU** | European Union |
| **FDA** | Food and Drug Administration |
| **FEW** | food, energy, water |
| **GC** | gas chromatography |
| **GHG** | greenhouse gas |
| **GWP** | global warming potential |
| **HAPs** | hazardous air pollutants |
| **HCFC** | hydrochlorofluorocarbons |
| **HCH** | hexachlorocyclohexane |
| **HFC** | hydrofluorocarbons |

| | |
|---|---|
| HVAC | heating, ventilation, and air conditioning |
| IHF | integrated horizontal flux |
| IPM | integrated pest management |
| LC | liquid chromatography |
| LD50 | lethal dose-50 |
| LOD | limit of detection |
| LOQ | limit of quantitation |
| LRT | long-range transport |
| MCPA | 2-methyl-4-chlorophenoxyacetic acid |
| MeBr | methyl bromide |
| MIC | methyl isocyanate |
| MITC | methyl isothiocyanate |
| MOE | margin of exposure |
| MS | mass spectrometry |
| MS/MS | tandem mass spectrometry |
| NCEI/NOAA | National Centers for Environmental Information/National Oceanic and Atmospheric Administration |
| NIOSH | National Institute for Occupational Safety and Health |
| NOEC | no observable effect concentration |
| NOEL | no-observed-effect level |
| NWF | National Wildlife Federation |
| NWS | National Weather Service |
| OA | oxygen analog |
| OC | organochlorine |
| ODP | ozone depleting potential |
| OEHHA | Office of Environmental Health Hazard Assessment |
| OP | organophosphates |
| OSHA | Occupational Safety and Health Administration |
| PAH | polycyclic aromatic hydrocarbons |
| PAN | peroxyacetylnitrates |
| PCBs | polychlorinated biphenyls |
| PCDD | polychlorinated dibenzo-$p$-dioxins |
| PFCA | perfluorinated carboxylic acids |
| PFOA | perfluorooctanoic acid |
| PFOS | perfluorooctanesulfonate |
| PM | particulate matter |
| POPs | persistent organic pollutants |
| PREC | Pesticide Registration and Evaluation Committee |
| PTFE | polytetrafluoroethylene |
| PUF | porous polyurethane foam |
| RAC | rotating arm collector |
| SPME | solid phase microextraction |
| SVOC | semivolatile organic compound |
| TAC | toxic air contaminant |
| TCDD | 2,3,7,8-tetrachlorodibenzo-$p$-dioxin |

| | |
|---|---|
| **TEPP** | tetraethyl pyrophosphate |
| **TFA** | trifluoroacetic acid |
| **TFM** | 3-trifluoromethyl-4-nitrophenol |
| **TLV** | threshold limit values |
| **TOCP** | tri-o-cresyl phosphate |
| **TPS** | theoretical profile shape |
| **TSCA** | U.S. EPA Toxic Substances Control Act |
| **TWA** | Trans-World Airlines |
| **UC, Davis** | University of California, Davis |
| **UNEP** | United Nations Environment Programme |
| **USDA-ARS** | United States Department of Agriculture–Agricultural Research Service |
| **USGS** | United States Geological Survey |
| **VOC** | Volatile Organic Compound |
| **WHO** | World Health Organization |

# Glossary of Agencies

**ACGIH (American Conference of Governmental Industrial Hygienists):** Charitable scientific organization that advances occupational and environmental health with science-based approaches. ACGIH establishes threshold limit values for chemical substances and physical agents and biological exposure indices.

**Cal EPA (California Environmental Protection Agency):** California's state agency that develops, implements, and enforces environmental laws that regulate air, water, and soil quality, pesticide use, and waste recycling and reduction.

**CDPR (California Department of Pesticide Regulation):** California's state agency that protects human health and the environment by regulating pesticide sales and use, and by fostering reduced-risk pest management. CDPR is the most comprehensive state pesticide regulation program in the United States.

**CARB (California Air Resources Board):** California's state agency that oversees all air pollution control efforts in California. CARB promotes and protects public health, welfare, and ecological resources through effective reduction of air pollutants while recognizing and considering effects on the economy.

**EPA (Environmental Protection Agency):** A U.S. federal agency tasked with environmental protection matters. EPA protects human health and the environment with federal laws that reduce environmental and human health risks, make sustainable communities, and work with nations to protect the global environment. For example, the Clean Air Act is the comprehensive federal law that regulates air emissions from stationary and mobile sources. Among other things, this law authorizes EPA to establish National Ambient Air Quality Standards to protect public health and public welfare and to regulate emissions of hazardous air pollutants.

**EU (European Union):** The E.U. is a political and economic union of 27 member states that are located primarily in Europe. The E.U. reviews air quality standards in line with the World Health Organization guidelines and provides support to local authorities to achieve cleaner air for its citizens.

**FDA (Food and Drug Administration):** A U.S. federal agency that protects the public health by ensuring the safety, efficacy, and security of human and veterinary drugs, biological products, and medical devices; and by ensuring the safety of the nation's food supply, cosmetics, and products that emit radiation.

**NOAA (National Oceanic and Atmospheric Administration):** A U.S. federal agency (part of U.S. Department of Commerce) that enriches life through science from the surface of the sun to the depths of the ocean floor; NOAA informs the public of the changing environment around them with daily weather forecasts, severe storm warnings, and climate monitoring to fisheries management, coastal restoration and supporting marine commerce.

**NWF (National Wildlife Federation):** The largest private, nonprofit conservation education and advocacy organization in the United States. The NWF protects wildlife and habitat while inspiring future generations of conservationists.

**NWS (National Weather Service):** A U.S. federal agency that provides weather forecasts, warnings of hazardous weather, and other weather-related products to organizations and the public for the purposes of protection, safety, and general information.

**OEHHA (Office of Environmental Health Hazard Assessment):** California's lead state agency for the assessment of health risks posed by environmental contaminants. OEHHA uses scientific evaluations that inform, support, and guide regulatory and other actions. OEHHA implements the Safe Drinking Water and Toxic Enforcement Act of 1986, commonly known as Proposition 65.

**OSHA (Occupational Safety and Health Administration):** A U.S. federal agency that ensures safe and healthful working conditions by setting and enforcing standards and by providing training, outreach, education, and assistance.

**Prop 65 (Safe Drinking Water and Toxic Enforcement Act of 1986):** Proposition 65 protects California's drinking water sources from being contaminated with chemicals known to cause cancer, birth defects, or other reproductive harm, and requires businesses to inform Californians about exposures to such chemicals. The current list, as of December 18, 2020 is at https://oehha.ca.gov/proposition-65/proposition-65-list.

**UNEP (United Nations Environment Programme):** UNEP is the leading global environmental authority. They assess global, regional, and national environmental conditions and trends; develop international and national environmental instruments; and strengthen institutions for the wise management of the environment in seven areas: climate change, disasters and conflicts, ecosystem management, environmental governance, chemicals and waste, resource efficiency, and environment.

**USDA-ARS (United States Department of Agriculture–Agricultural Research Service):** ARS is the USDA's chief in-house research agency. ARS delivers cutting-edge, scientific tools, and innovative solutions for American farmers, producers, industry, and communities to support the nourishment and well-being of all people; sustain the nation's agroecosystems and natural resources; and ensure the economic competitiveness and excellence of U.S. agriculture.

**USGS (United States Geological Survey):** The USGS is the sole science agency for the Department of the Interior. USGS provides science about natural hazards; water, energy, minerals, and other natural resources; the health of ecosystems and environment; and the impacts of climate and land-use change.

**WHO (World Health Organization):** WHO is an international agency that promotes health, keeps the world safe, and serves the vulnerable. WHO addresses: human capital across the life-course, noncommunicable diseases prevention, mental health promotion, climate change in small island developing states, antimicrobial resistance, and elimination and eradication of high-impact communicable diseases.

# Glossary of Terms

**Absorption:** The process of one material (absorbate) being retained by another (absorbent); this may be the physical solution of a gas, liquid, or solid in a liquid, or attachment of molecules of a gas, vapor, liquid, or dissolved substance to a solid surface by physical forces, etc. (Wikipedia)

**Adsorption:** The adhesion of atoms, ions, or molecules from a gas, liquid, or dissolved solid to a surface. This process creates a film of the adsorbate on the surface of the adsorbent. This process differs from absorption, in which a fluid is dissolved by or permeates a liquid or solid, respectively (https://unstats.un.org/unsd/environmentgl/default.asp).

**Aerosols:** A suspension of fine solid particles or liquid droplets in air or another gas. Examples of natural aerosols are fog, mist, dust, forest exudates, and geyser steam. Examples of anthropogenic aerosols are particulate air pollutants and smoke. System of solid or liquid particles suspended in a gaseous medium, having a negligible falling velocity. (https://unstats.un.org/unsd/environmentgl/default.asp)

**Atmosphere:** Mass of air surrounding the earth, composed largely of oxygen and nitrogen. (https://unstats.un.org/unsd/environmentgl/default.asp)

**Azeotrope:** If the total vapor pressure of aqueous mixtures deviates positively or negatively from Raoult's law, azeotropes can be formed at a particular mixture composition. A mixture of two or more liquids whose proportions cannot be altered or changed by simple distillation. When an azeotrope is boiled, the vapor has the same proportions of constituents as the unboiled mixture. (Wikipedia)

**Biopesticides:** Natural chemicals, or mixtures of natural chemicals, or chemicals based upon natural products that can be used to manage pests. The EPA defines biopesticides as "certain types of pesticides derived from such natural materials as animals, plants, bacteria, and certain minerals."

**Biota:** The living component of an ecosystem. (https://unstats.un.org/unsd/environmentgl/default.asp)

**Carbamates:** A newer class of cholinesterase inhibitors. They act by blocking nerve impulses and causing hyperactivity and tetanic paralysis of the insect, and then death. Most are not persistent and do not bioaccumulate in animals or have significant environmental impacts.

**Exposure:** The contact between an agent and the external boundary (exposure surface) of a receptor for a specific duration. Types of exposure include: aggregate exposure, which is combined exposure of a receptor to a specific agent from all sources across all routes and pathways, and cumulative exposure, which is total exposure to multiple agents that causes a common toxic effect(s) on human health by the same, or similar, sequence of major biochemical events. Human exposure to toxic chemicals that occurs via inhalation, ingestion, and dermal contact exposure, combined with potency, is used to estimate "risk." (https://www.epa.gov/sites/production/files/2020-01/documents/guidelines_for_human_exposure_assessment_final2019.pdf)

**Fugacity:** The "escaping tendency" of a chemical substance from a phase in terms of mass transfer coefficients.

**Hazardous Air Pollutants (HAPs):** Hazardous air pollutants, also known as toxic air pollutants or air toxics, are those pollutants that are known or suspected to cause cancer or other serious health effects, such as reproductive effects or birth defects, or adverse environmental effects. HAPs are regulated by the United States EPA. Examples of HAPs are, benzene, which is found in gasoline; perchloroethylene, which is emitted from some dry-cleaning facilities; methylene chloride, which is used as a solvent and paint stripper by a number of industries; dioxin; asbestos; toluene; and metals such as cadmium, mercury, chromium, and lead. (https://www.epa.gov/haps/what-are-hazardous-air-pollutants)

**LD50:** The threshold level of exposure to toxic substances beyond which 50% of a population or organisms cannot survive. (https://unstats.un.org/unsd/environmentgl/default.asp)

**Long-Range Transport:** The atmospheric transport of air pollutants within a moving air mass for a distance greater than 100 km. (https://unstats.un.org/unsd/environmentgl/default.asp)

**Mesocosm:** Any outdoor experimental system that examines the natural environment under controlled conditions. Mesocosm studies may be conducted in an enclosure or partial enclosure that is small enough so that key variables can be brought under control.

**Microcosm:** Enclosed laboratory chamber systems that are constructed to simulate natural systems on a reduced scale, and for measuring responses to varying conditions (e.g., moisture, nutrients, sunlight, and temperature) over time.

**Neonicotinoids:** Insecticides chemically related to nicotine. They act on nerve synapse receptors and are much more toxic to invertebrates than to mammals, birds, and other higher organisms. (https://citybugs.tamu.edu/factsheets/ipm/what-is-a-neonicotinoid/)

**Organochlorine (OC) pesticides:** Chlorine containing pesticides that control a wide range of insects, including malaria-bearing mosquitoes, pests in food and fiber crops, and pests of livestock. Characterized by their low polarity and persistence in soil, aquatic sediments, and air, they tend to bioconcentrate in the tissues of invertebrates and vertebrates from their food.

**Organophosphate (OP) insecticides:** They act by inhibiting the enzyme acetylcholinesterase. Most are not persistent and do not bioaccumulate in animals or have significant long-term environmental impacts. OPs have a wide range of toxicities; some are neurotoxic.

**Particulate matter:** Fine liquid or solid particles, such as dust, smoke, mist, fumes, or smog, found in air or emissions. Suspended particulate matter is finely divided solids or liquids that may be dispersed through the air from combustion processes, industrial activities, or natural sources. (https://unstats.un.org/unsd/environmentgl/default.asp)

**Pheromones:** A pheromone is a secreted or excreted chemical factor that triggers a social response in members of the same species. (Wikipedia)

**Semiochemicals:** Chemicals involved in the biological communications between individual organisms. Includes allelochemicals (chemicals produced by a living organism exerting a detrimental physiological effect on the individuals of another species when released into the environment) and pheromones. (Wikipedia)

**Semi-volatile organic compounds (SVOCs):** A subgroup of volatile organic compounds (VOCs) that tend to have a higher molecular weight and higher boiling point temperature than other VOCs. SVOCs include pesticides (DDT, chlordane, plasticizers (phthalates), and fire retardants (PCBs, PBB)). SVOCs have vapor pressures between $10^{-5}$ and 0.1 Pa.

**Stratosphere:** The upper layer of the atmosphere (above the troposphere), between approximately 10 and 50 km above the earth's surface. (https://unstats.un.org/unsd/environmentgl/default.asp)

**Toxic air contaminant (TAC):** An air pollutant which may cause or contribute to an increase in mortality or an increase in serious illness, or which may pose a present or potential hazard to human health. (CARB: California Air Resources Board)

**Troposphere:** The layer of the atmosphere extending about 10 km above the earth's surface. (https://unstats.un.org/unsd/environmentgl/default.asp)

**Vapor:** The gas phase of a substance at a temperature where the same substance can also exist in the liquid or solid state, below the critical temperature of the substance. (Wikipedia)

**Volatile organic compounds (VOCs):** Organic compounds that evaporate readily and contribute to air pollution mainly through the production of photochemical oxidants. Some VOCs are formaldehyde, toluene, acetone, ethanol, isopropyl alcohol, and hexane. (https://unstats.un.org/unsd/environmentgl/default.asp)

# 1

# Introduction: A Summary with New Perspectives

## 1.1 Introduction

Contaminants in air are responsible for more premature deaths worldwide than those in any other environmental media. The topic of air has never been more important in my (JNS) lifetime than it is now. In 2020, the world came to a virtual halt due to the novel coronavirus (COVID-19), an airborne virus that caused a pandemic. As of June 2021, 170 million people worldwide had been infected, and 3.5 million people died (600,000 people in the United States alone). As with other airborne contaminants, steps can be taken to lessen the effects. The first measure in many places was to stay home. And in the beginning as people shuttered up in their homes, animals took over abandoned cities and swimming pools became duck ponds. One potential upside of people staying home during the pandemic has been a sharp decrease in greenhouse gases (Plumer, 2021).

But much of the world has tried to return to daily life (whether or not it is premature is a contentious subject). To slow the spread, many strategies have been used in combination: masks that can filter virus-laden air, social distancing, vaccines, and virus-sniffing dogs. Although the vaccines were released with unprecedented speed and some are nearly 100% effective, the vaccination programs have been overwhelmed with logistical problems due to the challenge of vaccinating the entire population, vaccine shortages in some regions of the world, and vaccine hesitancy in some populations.

Also in 2020, weather and natural disaster events (due to global climate change) increased in duration and magnitude. California and the West Coast have experienced the worst wildfires in U.S. history, with more than 4 million acres of urban, farmland and forests being destroyed. Among the worst wildfires in California history, five were in 2020, and four of those were in August alone (CAL FIRE, 2021). Wildfires release particulate matter, $CO_2$, and incomplete combustion products that are known carcinogens (e.g., polycyclic aromatic hydrocarbons (PAHs)). And when pesticides and other applied chemicals are present, they are released to air.

DOI: 10.1201/9781003217602-1

It is urgent in this landscape to think and progress in a different way. Encouragingly, the new U.S. administration, sworn in January of 2021, is reinstating the United Sates into the Paris Agreement (an international effort to curb the generation of greenhouse gases) and banning new oil pipelines in Federal lands, signaling a renewed interest to improve the quality of air.

The implementation of FEW (food, energy, water) Nexus sustainability guidelines has progressed worldwide. The premise of the FEW Nexus is to consider the effects a technology developed for one sector will have on the other sectors. For example, a new farming technique that promises to feed more people must also consider the impacts it will have on water and energy usage. In line with FEW Nexus principles, biopesticides and more robust crops (both genetically engineered and traditionally bred) show promise for future reductions in the environmental impact of agriculture. Most of the new pesticides registered in the United States are biopesticides, and China has set policies to encourage development and adoption of biopesticides and sustainable practices in its vast farming activities (Figure 1.1). However, in order to feed the world's growing population, especially those in economically developing nations like China, it is projected that use of synthetic pesticides will increase substantially before biopesticides and complementary farming practices that reduce the use of synthetic pesticides become the norm.

This book is a compendium of over 30 years of research addressing contaminants in the air. Our work initiated and paved the way for being able to examine the quality of ambient air by identifying and quantifying its contaminants, leading to comprehensive understanding of transport, fate, and exposure of pesticides and toxics in air. It is a product of over three decades of research, teaching, and outreach by our group and our cohorts at the University of California, Davis; the University of Nevada, Reno; the United States Department of Agriculture–Agricultural Research Service (USDA-ARS); and Arizona State University dating roughly from 1970 to 2010. Most of the original research was in collaboration with scientists and agency personnel and has subsequently been published in peer-reviewed papers, reviews, book chapters, dissertations, reports, and forums such as American Chemical Society (ACS) symposia and other national and international venues.

---

## 1.2 Distribution and Fate of Pesticides in Air

It all started with a simple question: where do all the chemicals released into the environment go? Some argued for water drainage to streams, ponds, lakes, and oceans. Others for soil breakdown. Yet no one knew for certain. We postulated air and transformation to be the missing elements. Chemicals emitted directly to air, or those that volatilize slowly into air, break down and become diluted in the huge volume of air and "disappear." Simple, right?

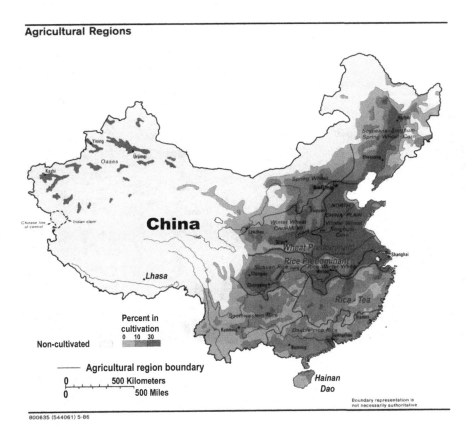

**FIGURE 1.1**
Agricultural regions of China. (Image credit: U.S. Central Intelligence Agency.)

Not hardly. The complexity is evident when one views all the possibilities (Figure 1.2).

Among the pollution caused by the industrial revolution, air pollution was most striking. Shift from coal and regulations put into place in the developed world have drastically improved air quality in developed countries. As discussed in Chapter 4, regulations by the U.S. Environmental Protection Agency (EPA) and various state and local municipalities have improved the quality of air in the United States (similar measures were successful in other developed regions as well) (Landrigan et al., 2018). While conditions in developed regions are improving, rapidly industrializing nations today are experiencing decreasing air quality (Landrigan et al., 2018). Initiatives to shift biomass burning to biofuel production in developing nations, especially in Asia, may change this in the coming years. In 2016, of the 9 million

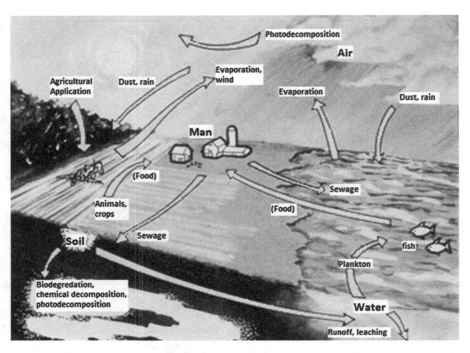

**FIGURE 1.2**
Pesticide cycling in the environment. (Reproduced from Author Unknown (1974) Scientists probe pesticide dynamics. *Chem. Eng. News Archive 52*, 32–33. Copyright (1974) American Chemical Society).

premature deaths caused by pollution, 6.5 million are estimated to be from airborne pollution.

Pesticides have been intentionally released to the environment for economic reasons at higher rates than any other class of chemicals (Woodrow et al., 2018). There is much we do not know about the biological consequences of low levels of pesticides and other toxics in ambient air. Our research, plus the research of others, over the past 30 years has made it clear that the atmosphere is not an "infinite reservoir" and that pesticides and other toxics released to air do not simply go away (Woodrow et al., 2018). As described in depth in Chapter 10, long-range transport, as measured by transect studies, occurs with pesticides. While stable and nonpolar pesticides might be found in low levels in air and water in remote locations, they have a tendency to bioaccumulate to high levels, especially in higher trophic level organisms. Until about that time (30 years ago), the fate of pesticides released to the environment was not considered for their impact on human and ecosystem health. The tipping point was the public awareness brought about by Rachel Carson's book, *Silent Spring* in 1962 (Carson, 1962). Eventually, it led to the ban of DDT and other pesticides and the creation of the U.S. EPA in 1970.

We all experience air in different ways—as a means to sustain and refresh ourselves, to catch our breath, and to sense aromas and odors in our surroundings. Air is also the medium through which we see our surroundings. We inhale and exhale on average a liter per minute which can expose us to air's contents. There is more to air besides oxygen, nitrogen, and carbon dioxide. Think of fresh air compared to polluted air, containing odors from perfumes, freshly baked bread, refineries, chemical complexes, auto exhaust, skunks, and cattle feedlots. This bouquet of odors is due to a complex chemical mixture. New chemists in industry are told the odor from manufacturing was the smell of money! Over time they realized it was pollution and potentially harmful.

Odor is not, however, the only indicator of chemicals in air. Think also about insect pheromones that are emitted in femtogram quantities to attract male moths to females and the spray from a crop duster that kills pests or the lethal Great Smog events during early industrialization.

Air is also a transport and exposure route for actual and potential exposures to people, e.g., within the current pandemic of COVID-19, valley fever, and other human diseases, and for exposures to wildlife: birds, bees, and rare and endangered species globally. Air is a conduit for harmful chemicals and diseases.

In this book, we focus on potentially harmful chemicals in air with an emphasis on volatile and semivolatile pesticides, related contaminants both synthetic and naturally occurring, as well as viruses and pathogens. These issues will be discussed in relation to their physicochemical properties and how that affects their environmental fate and distribution, and ways to sample and analyze them, as well as opportunities to use this information to minimize risk. We will touch on computer-based modeling related to emissions and dispersion of pesticides and other contaminants in air. The ambient monitoring of air for residues will be briefly discussed, particularly in relation to opportunities to improve data quality and their use in risk assessment (Brooks, 2012; Crosby, 1998).

The topic is of interest to politicians, regulators, and environmental scientists, including students, ecologists, and public health officials dealing with smoke from fires one day and airborne viruses the next. And it can have practical consequences in explaining tragic accidents, like the fatal crash of a TWA airliner on Long Island, and deteriorating air quality in the valleys and air basins of the West. It can answer concerns over wildlife—why hawks and other valuable species become sick and die from exposures in orchards. And why some species, amphibians and others, are disappearing from mountainous terrains, perhaps due to the presence of endocrine system disruptors that are residues from synthetic chemicals.

"The more we know, the more we grow"—an old saying that in this context can help us understand and control our impact on the air environment. It's all about chemicals in air!

## References

Brooks, L. (2012). *Department of Pesticide Regulation Air Monitoring Shows Pesticides Well Below Health Screening Levels.* Sacramento: California Department of Pesticide Regulation.

CAL FIRE (2021). https://www.fire.ca.gov/ (Accessed January 29, 2021).

Carson, R. (1962). *Silent Spring.* Boston, MA: Houghton Mifflin.

Crosby, D.G. (1998). *Environmental Toxicology and Chemistry.* New York: Oxford University Press.

Landrigan, P.J., Fuller, R., Acosta, N.J.R., Adeyi, O., Arnold, R., Baldé, A.B., et al. (2018). The Lancet commission on pollution and health. *Lancet* 391, 462–512.

Plumer, B. (2021). Covid-19 took a bite from U.S. greenhouse gas emissions in 2020. *N. Y. Times.*

Woodrow, J.E., Gibson, K.A., and Seiber, J.N. (2018). Pesticides and related toxicants in the atmosphere. In P. de Voogt (Ed.), *Reviews of Environmental Contamination and Toxicology* (Vol. 247, pp. 147–196). Cham: Springer International Publishing.

Author Unknown (1974). Scientists probe pesticide dynamics. *Chem. Eng. News Archives* 52, 32–33.

# 2

# Historical and Current Uses of Pesticides

## 2.1 Introduction

A well-known song by Huddie Ledbetter in 1940 chronicles the cotton bolls too rotten to allow harvest—very appropriate lyrics to the saga of pesticide use on cotton, a major use over many years in the southern United States and other parts of the world.

Pesticides are single chemicals or mixtures used by humans to restrict or repel pests such as bacteria, nematodes, insects, mites, mollusks, birds, rodents, weeds, and other organisms that affect food production or human health. They disrupt some component of the pest's life processes to kill or inactivate it.

## 2.2 Early Pesticides

Before the 1880s, very few chemicals were used for pest control. Most of these early pest control substances were inorganics or botanicals. Some examples include sulfur, copper sulfate, and pyrethrin extracts. Lead arsenate was heavily used on apples to control codling moths; the common practice was to wash it off before eating. We still find elevated lead and arsenic residues in agricultural fields, playground, and subdivision soils—places where they are not expected.

Following are some of the early discoveries, toxic symptoms, and applications of pesticides (Ware, 1983):

- **1882—White arsenic**

  Widespread use as a contact herbicide from 1900 to 1910

  Ingestion causes vomiting, abdominal pains, diarrhea, and bleeding. Sublethal doses can lead to convulsions, cardiovascular distress, liver and kidney inflammation, and abnormal blood coagulation. It can be lethal.

- **1891—Lead arsenate**

  Very effective insecticide that clings to the plant longer

DOI: 10.1201/9781003217602-2

Heavily used on fruit orchards, including apples to control codling moths

Most alternatives were less effective or more toxic

Persistent in the environment

- **1892—Dinitrophenol**

  Used as an herbicide, insecticide, and fungicide

  Very toxic uncoupler of oxidative phosphorylation, preventing formation of adenosine triphosphate

- **1897—Oil of Citronella**

  A natural insecticide from citronella grass

  Can be used topically by humans to control mosquitos

- **1904—Potassium cyanide**

  A broad-spectrum fumigant (after reaction with acid to form hydrogen cyanide gas) for protecting plants from insects

  Very toxic uncoupler of oxidative metabolism

- **1908—Nicotine**

  Contact insecticide and fumigant, mostly effective against soft-bodied insects, or those of small size (mites, thrips, aphids)

  Led eventually to development of neonicotinoids in the 1980s–1990s

  Disrupts nervous system—toxic to animals including humans

In the 1920s, arsenic-based insecticides became the predominant pesticide used around the world. In the United States alone, calcium arsenate was used in huge amounts to control the boll weevil. In 1919, 1.5 million kg were used, and the usage jumped to 5 million kg in 1920. Aerial spraying from planes became common in the 1920s. Around 1925, some experts began to raise concerns about residues. In 1926, thallium sulfate, which is toxic to humans, was introduced to control ants and rats. In the 1930s, rotenone, a plant root extract, came into use as an insecticide; it proved to be poisonous to fish though not to humans.

---

## 2.3 The Synthetic Revolution

### 2.3.1 Organochlorines

The next generation of pest control was based on synthetic chemicals called organochlorines (OCs). OCs of generally low to moderate toxicity were introduced during and after World War II. The best-known example is DDT (dichlorodiphenyltrichloroethane), which was introduced in 1939 by Herman Paul Müller, a Swiss chemist.

After its introduction, use of DDT expanded rapidly. DDT was cheap to make and was a good control for malaria-bearing mosquitoes. The lethal dose-50 ($LD_{50}$) for DDT to rats is 113 mg/kg—thus it was considered generally safe for applicators. Other OCs introduced in the 1940s and 1950s included benzene hexachloride, chlordane, toxaphene, aldrin, dieldrin, endrin, endo-sulfan, isobenzan, and methoxychlor. Methoxychlor was even less toxic than DDT and broke down more readily in the environment. Despite this, DDT use dominated.

These OCs were used to control a wide range of insects, including malaria-bearing mosquitoes, pests in food and fiber crops, and pests of live-stock. Overuse led to resistant pests and toxicity to nontarget organisms: the "pesticide treadmill." Their low polarity and persistence in soil, aquatic sediments, air, and biota caused a tendency to bioconcentrate in the tissues of invertebrates and vertebrates. The OCs could move up trophic chains and affect top predators (e.g., eagles and salmon). Industry started withdrawing production of OCs including DDT in the 1970s over carcinogenicity concerns and damage to wildlife (Figure 2.1). The popular book *Silent Spring* (Carson, 1962) focused attention on the risks surrounding use of OCs.

## 2.3.2 Organophosphates

In March 1968, thousands of sheep died suddenly and unexpectedly over-night in Skull Valley, Utah. Over the next few weeks, other sheep fell ill and, finally, a total of 6,000 sheep died (Boissoneault, 2018). Their symptoms were

**FIGURE 2.1**
Waste DDT at the Montrose chemical site near Palos Verde, CA. Although the Montrose site is now closed, remnants of the outflow from Montrose have contaminated fisheries along the Palos Verde shelf. Fishermen are advised not to eat fish from this area. (Image credit: National Oceanic and Atmospheric Administration (NOAA).)

characteristic of nerve gas: the sheep were dazed, heads tilted to one side, and movements were uncoordinated (Boffey, 1968).

Nearby residents suspected the Dugway Proving Ground, a U.S. military chemical and biological weapons research and testing facility, which was only 27 miles away. The military's role in the incident was confirmed a week later when a report was released stating that a high-speed jet had malfunctioned during a test and sprayed 320 gallons of nerve gas Venomous Agent X (VX) at a much higher altitude than expected (Boissoneault, 2018).

Most Western countries banned the use of chemical weapons following the First World War, which incurred 90,000 deaths and 1 million injuries due to chemical weapons; however, the United States did not enter into the agreement. The United States officially ended its use, development, and stockpiling of chemical weapons in 1997 when a treaty signed at the *Convention on the Prohibition of the Development, Production, Stockpiling and Use of Chemical Weapons and on their Destruction* came into full effect (U.N. Office for Disarmament Affairs, 2021). VX is a semivolatile (vapor pressure=0.09 Pa) organophosphorus chemical similar in structure and mode of action to organophosphate (OP) insecticides.

OP insecticides originated from compounds developed as nerve gases during World War II. The nerve gas-inspired insecticides, such as tetraethyl pyrophosphate (TEPP) and parathion, have high mammalian toxicities ($LD_{50} < 20$ mg/kg). They act by inhibiting the enzyme acetylcholinesterase (AChE) (shown in Figure 2.2) that breaks down the neurotransmitter acetylcholine (ACh) at the nerve synapse, blocking impulses and causing hyperactivity and tetanic paralysis of the insect, and finally death. Most are not persistent and do not bioaccumulate in animals or have significant long-term environmental impacts.

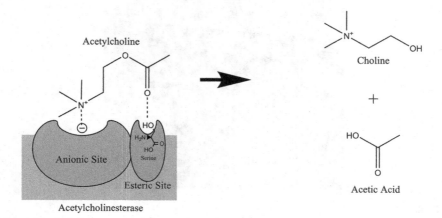

**FIGURE 2.2**
Mode of action of OPs and also carbamates is inhibition of the nerve signal via the vital enzyme AChE. (Adapted from Wiener and Hoffman, 2004.)

In addition to acting as AChE inhibitors, some OPs can cause a delayed effect that may not be reversible. Several non-insecticidal OPs, including the widely used plasticizer and gasoline additive tri-o-cresyl phosphate (TOCP), cause a longer lasting and more insidious poisoning effect termed OP-induced delayed neuropathy. This delayed axonal neuropathy results in sensory effects, slow degeneration of leg muscle control, and eventually paralysis. During prohibition, thousands of people experienced partial paralysis after unintentionally consuming TOCP that was mistaken for an alcoholic beverage—termed "ginger jake" paralysis. Additional poisoning resulted from the use of TOCP to dilute cooking oil. In all cases, the poisoning agent was not TOCP per se, but rather a dioxaphosphorin metabolite of TOCP (Figure 2.3).

The toxicity of the pesticide leptophos (Phosvel) was related to this (Figure 2.4). This chemical had short-lived use as an insecticide on cotton. After its delayed neurotoxic effects showed up in farm animals in the Middle East, it was banned from commercial use as a pesticide.

Author JNS worked with pesticides related to leptophos, and occasionally with TOCP derivatives. He now has partial leg paralysis with symptoms reminiscent of delayed neuropathy. While there may be a connection, it is

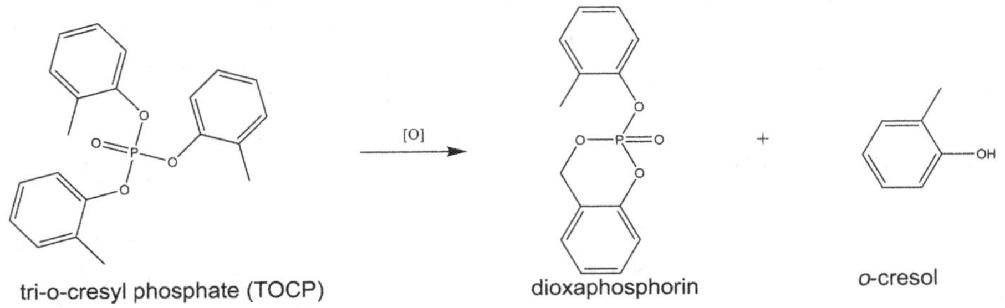

**FIGURE 2.3**
Oxidation of TOCP to *o*-cresol and its toxic metabolite dioxaphosphorin. (Adapted from Casida et al. 1961; Crosby 1998.)

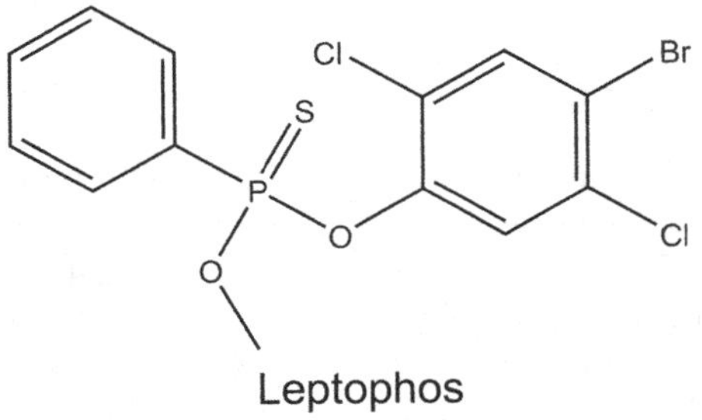

**FIGURE 2.4**
Leptophos was banned from commercial use as a pesticide.

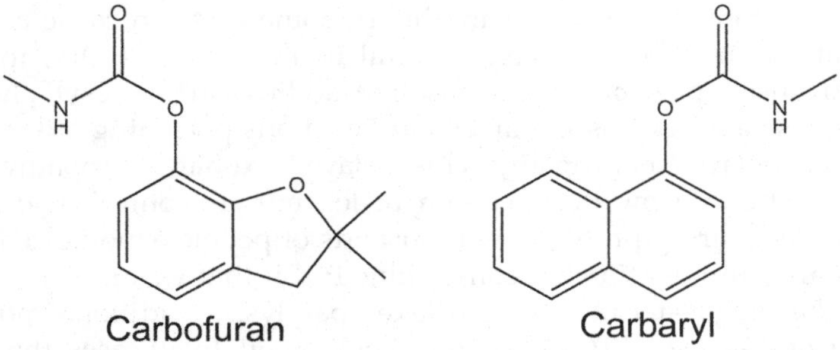

**FIGURE 2.5**
Carbamates include carbofuran and carbaryl.

hard to know since the exposure took place 50 years or more before paralytic effects were noticed (see the discussion of paraquat toxicity in Chapter 4).

### 2.3.3 Carbamates

Carbamates are a newer class of cholinesterase inhibitors. They act by blocking nerve impulses of insects, causing hyperactivity, tetanic paralysis, and then death. Most are not persistent and do not bioaccumulate in animals or have significant environmental impacts. Carbamate insecticides include, among others, aminocarb, fenobucarb, aldicarb, carbofuran (Figure 2.5), fenoxycarb, and carbaryl (Figure 2.5). They are generally less toxic to mammals than OPs but are toxic to bees.

Carbaryl was introduced by Union Carbide in about 1950 and is still used today in homes and on crops. It is of low toxicity and moderate persistence. Methomyl and related carbamates like Temik are effective low residual carbamates but have potential to leach in sandy soils and contaminate groundwater. The unorthodox use of N-methyl carbamates to hasten the growth of cucumbers and melons has led to sporadic reversible symptoms of AChE inhibition in consumers.

### 2.3.4 Synthetic Herbicides

Popular synthetic herbicides began with the introduction of phenoxy herbicides 2-methyl-4-chlorophenoxyacetic acid (MCPA), 2,4-dichlorophenoxyacetic acid (2,4-D), and 2,4,5-trichlorophenoxyacetic acid (2,4,5-T) (Figure 2.6). The latter two are components of "Agent Orange" used by the U.S. military as a defoliant during the Vietnam War. Agent Orange is on the "dirty dozen" list of the World Health Organization as off-limits for nations worldwide because it is contaminated with 2,3,7,8-tetrachlorodibenzo-p-dioxin (TCDD; "dioxin")—one of the most toxic synthetic chemicals known to humans

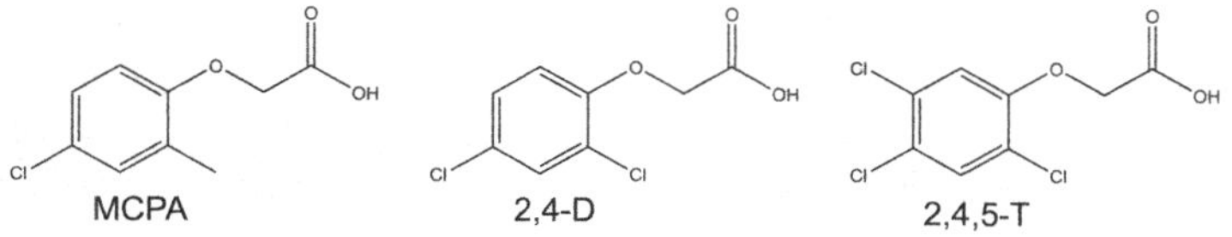

**FIGURE 2.6**
Synthetic herbicides include MCPA, 2,4-D, and 2,4,5-T.

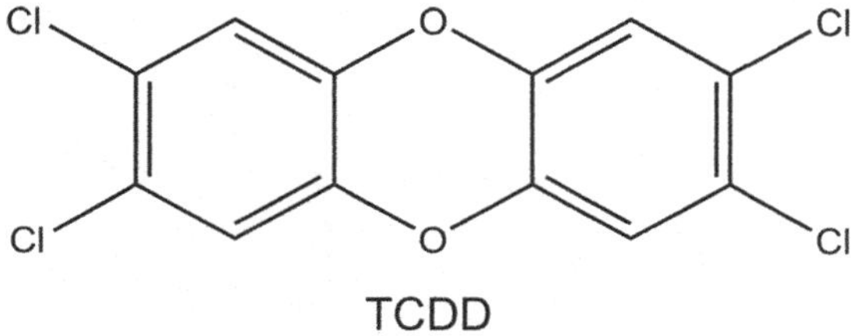

**FIGURE 2.7**
TCDD is one of the most toxic chemicals known to humans.

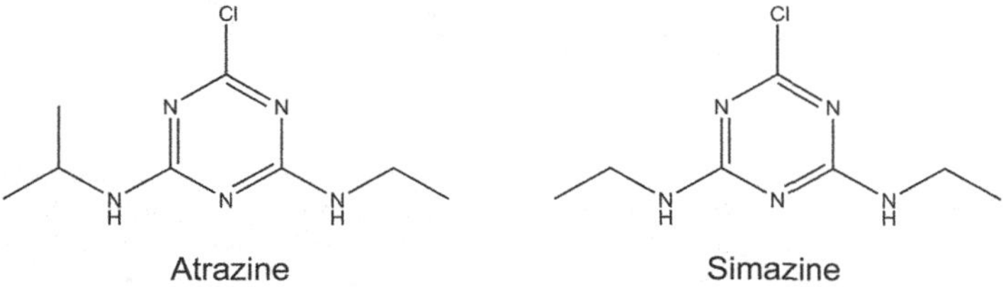

**FIGURE 2.8**
Triazine herbicides include atrazine and simazine.

(Figure 2.7). TCDD is still seen in air samples collected near municipal dump sites and particularly around incinerators (Shibamoto et al., 2007).

The triazine herbicides, atrazine and simazine, are heavily used on row crops such as corn in the United States and elsewhere resulting in contamination of surface water and groundwater, and frequently are found in ambient air samples (Figure 2.8).

Glyphosate or Roundup™ (Figure 2.9) has become the dominant herbicide in the United States since its introduction by Monsanto in about 1980. Glyphosate is a broad-spectrum systemic herbicide that leaves no significant toxic residues and breaks down in the environment to 2-amino-3-(3-hydroxy-5-methyl-isoxazol-4-yl)propanoic acid (AMPA). It can often be used without a permit and is of relatively low toxicity to people and wildlife. However, overuse can kill desirable vegetation such as milkweeds, *Senecio* species (e.g., ragwort and groundsel), and other deep-rooted perennials that serve as refuge, food, or nectar sources for desirable species like

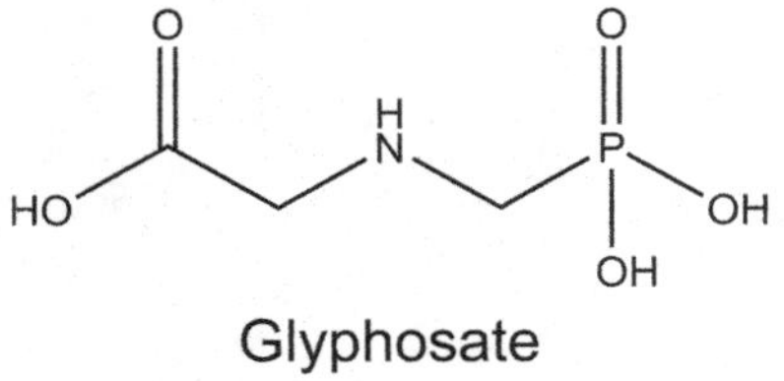

Glyphosate

**FIGURE 2.9**
Glyphosate or Roundup™ is an herbicide.

**FIGURE 2.10**
Loss of habitat for monarch butterflies due to overuse of glyphosate. Milkweeds are required host plants for monarchs. Glyphosate kills deep-rooted perennials like milkweeds. Overuse of glyphosate or glyphosate drift will eliminate the host plant milkweeds resulting in fewer adult monarchs. Seen as a reduced number at overwintering sites. (Photo credit: Margaret Baker.)

butterflies (Figure 2.10). Overuse can occur during control of weeds in glyphosate-resistant crops like soybeans. As with many herbicides, glyphosate drift can harm nontarget vegetation, and there are reports of chronic toxicity among users, resulting in litigation which is ongoing.

## 2.3.5 Neonicotinoids

Neonicotinoids are promising new chemical insecticides, but their future is clouded by allegations that they are toxic to bees under conditions of normal use per label instructions. Adverse ecological effects include the honeybee colony collapse disorder and loss of birds due to a reduction in insect populations. Since 2013, several countries have restricted the use of neonicotinoids such as imidacloprid, clothianidin, and thiamethoxam (Figure 2.11) (McDonald-Gibson, 2013).

The current status of neonicotinoids use and regulatory status in California is given in some detail below to illustrate the labyrinth of evaluations pesticides must go through (Box 2.1).

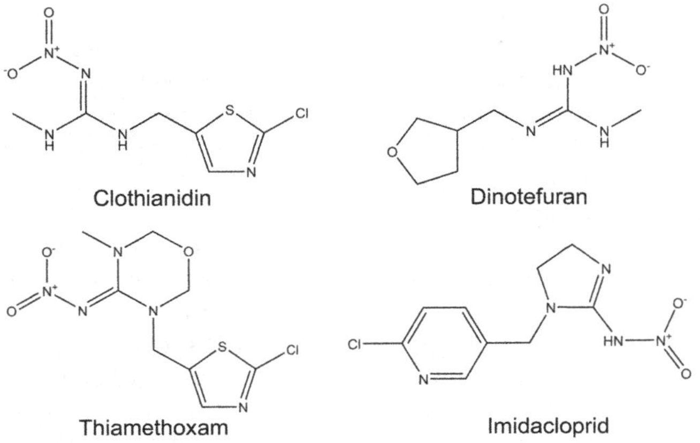

**FIGURE 2.11**
Neonicotinoids include imidacloprid, clothianidin, thiamethoxam, and dinotefuran.

## BOX 2.1  OVERVIEW OF CALIFORNIA DEPARTMENT OF PESTICIDE REGULATION'S (DPR) NEONICOTINOID REEVALUATION (DARLING AND CLENDENIN 2020)

### PESTICIDE REGISTRATION AND EVALUATION COMMITTEE (PREC) MEETING MINUTES—JULY 17, 2020

Over the last several years, honeybee colony decline has triggered worldwide concern. The most common factors involved in colony decline include predatory insects, malnutrition, genetic diversity in queen bees, and direct or indirect pesticide exposure. These factors can have complex interactions and cause compounding effects, however, because DPR has regulatory authority over pesticides, the department's efforts are focused strictly on the interaction between pesticides and pollinators.

Several categories of pesticides may be affecting bees, including fungicides, insecticides, tank mixtures, or a combination thereof. DPR received early adverse effects reports identifying possible risk to bees from exposure to imidacloprid (Figure 2.11). This active ingredient is part of a class of chemicals called nitroguanidine-substituted neonicotinoids (neonics), which also include clothianidin, dinotefuran, and thiamethoxam (Figure 2.11).

Neonics are systemic pesticides, meaning they have the ability to translocate throughout the plant once absorbed through the roots or foliage. These pesticides can be applied as a foliar or soil application and are especially effective on sucking insects such as aphids. Neonics are registered for use on a wide variety of crops worldwide including citrus, pome fruit, stone fruit, cotton, and cereal grains. They were developed as an alternative to OPs and carbamates. Imidacloprid was first registered in California in 1994, while the other three active ingredients were registered in 2004.

A reevaluation can be triggered in a variety of different ways, such as issues with phytotoxicity or efficacy, or concerns over impacts on worker health or the environment. As mentioned previously, the neonic reevaluation was triggered by adverse effects data submitted by the registrant, which showed high levels of imidacloprid in both leaves and blossoms of treated plants. These levels were high enough to be expected to be hazardous to bees. Based on DPR's evaluation of the adverse effects data, DPR issued California Notice 2009-02, which formally placed the four neonic active ingredients into reevaluation in February 2009. The reevaluation focuses on outdoor agricultural uses of neonics and does not include products that are not relevant to the concerns that prompted the reevaluation. This means that products such as ant and cockroach baits or indoor use products were not pulled into the reevaluation. For more information on the specifics of the initiation of this reevaluation, contact CDPR or visit DPR's Neonicotinoid Reevaluation Web page <https://www.cdpr.ca.gov/docs/registration/reevaluation/chemicals/neonicotinoids.htm>.

After DPR initiated the reevaluation, staff began working collaboratively with U.S. EPA and the Health Canada Pest Management Regulatory Agency to determine data requirements and develop study designs. DPR's neonic reevaluation relies on data submitted by the registrants and peer-reviewed open literature studies. The specific data required included honeybee toxicity studies, colony feeding studies, and pollen nectar residue studies. DPR worked with U.S. EPA on the preliminary pollinator risk assessments, looking at acute toxicity endpoints as well as preliminary residue studies and colony feeding studies. U.S. EPA issued the preliminary risk assessment for imidacloprid in January 2016 and for clothianidin, thiamethoxam, and dinotefuran in January 2017.

In July 2018, DPR published the California Neonicotinoid Risk Determination. For this risk determination, DPR scientists assessed data from residue trials of the four neonicotinoids at various rates and timings, and compared them to colony feeding studies. The data from the colony feeding studies was used to determine residue levels that pose no significant toxicity to bees. The residue studies assessed in the document were conducted under worst-case scenarios, based on the maximum application rates and minimum reapplication intervals listed on California labels over a 2-year period. Throughout the review of the data, DPR scientists identified risks to bees when using neonics on a number of crops. After publishing the risk determination document, DPR received additional information that was then incorporated into an addendum to the risk determination and published in January 2019. These documents were used to inform DPR's risk mitigation efforts. For more information regarding the scientific methods that were used in producing this risk determination, view the risk determination document <https://www.cdpr.ca.gov/docs/registration/reevaluation/chemicals/neonicotinoid_risk_determination.pdf>.

Since publishing the January 2019 addendum to the risk determination, DPR has been working on how to appropriately mitigate the identified risks. Staff have pursued finding a complex balance between the need to keep a critical crop protection tool available, while simultaneously trying to understand current cultivation practices, pest pressures, and use patterns. DPR used that understanding to develop proposed mitigation measures to protect bees. In addition to the worst-case scenario residue studies, DPR has additional studies on file that were not based on worst-case scenarios. In the time since publishing the addendum, DPR has reviewed these studies to identify whether lower application rates could serve as a mitigation option.

DPR is proposing to implement the needed mitigation measures by adopting regulations, rather than through label changes. The draft regulations include both general mitigation measures and specific mitigation by crop group. The uses of neonics pose different risks depending on the crop to which they are applied, and in some cases, depending on which neonic is used.

Additional information will be sent out through DPR's California Notice to Stakeholders electronic mailing list. Following the webinars, DPR will have an open 30-day comment period to solicit comments from stakeholders. DPR will consider all feedback and make adjustments to the mitigation where appropriate. The goal is to develop the final rulemaking package and begin the official public notice process later in 2020.

While DPR progresses through the rulemaking process for neonics, the department has already instituted several bee protections. The California Managed Pollinator Protection Plan is a central location document to find all of California's initiatives to protect pollinators, including current laws and regulations, citrus bee protection areas, bee protection practice agreements, bee registration and notice of applications, and current label language.

In January 2020, U.S. EPA released the Preliminary Interim Decision for each of the neonic active ingredients. U.S. EPA conducted a broader risk assessment, which assessed impacts to pollinators, human health, aquatic risks, and ecological risks. Although DPR is preempted from requiring changes to pesticide product labels, U.S. EPA proposed several label changes in their Preliminary Interim Decision to protect bees and address other risks. The comment period for this document closed on June 18, 2020, and the next steps are to review the comments and publish a final decision.

## 2.4 Biopesticides

Biopesticides are the newest generation of pesticides (Seiber et al., 2014). The category of all "pesticides" are chemicals, including mixtures of chemicals, used by humans to restrict or repel pests, such as bacteria, nematodes, insects, mites, mollusks, birds, rodents, and other organisms, that adversely affect food production or human health. The subcategory of "biopesticides" includes natural chemicals, or mixtures of natural chemicals, which are based upon natural products and that manage pests. The EPA defines biopesticides as "certain types of pesticides derived from such natural materials as animals, plants, bacteria, and certain minerals."

Biochemical pesticides are naturally occurring substances that control pests by nontoxic mechanisms. Conventional pesticides, by contrast, are generally synthetic materials that directly kill or inactivate the pest. Biochemical pesticides include substances that interfere with mating, such as insect sex pheromones, as well as various scented plant extracts that attract insect pests to traps. The EPA notes that "As of April 2016, there are 299 registered biopesticide active ingredients and 1401 active biopesticide product registrations."

Important qualities sought out in biopesticides include target selectivity, lacking undue persistence, and low or minimal nontarget toxicity. The EPA notes the following "advantages of using biopesticides":

- Biopesticides are usually inherently less toxic than conventional pesticides.
- Biopesticides generally affect only the target pest and closely related organisms, in contrast to broad spectrum, conventional pesticides that may affect organisms as different as birds, insects, and mammals.
- Biopesticides often are effective in very small quantities and often decompose quickly, resulting in lower exposures and largely avoiding the pollution problems caused by conventional pesticides.

- When used as a component of integrated pest management (IPM) programs, biopesticides can greatly reduce the use of conventional pesticides, while crop yields remain high (*US EPA*).

### 2.4.1 Spinosad

Spinosad is a biologically derived insecticide and a leading biopesticide. It is produced by the fermentation culture of *Saccharopolyspora spinosa*, which is a soil organism. Spinosad is a mixture of two macrocyclic lactones in a tetracyclic ring (Figure 2.12, Table 2.1). Each component is an unsaturated tetracyclic ester with two sugar derivatives, forosamine and rhamnose. Spinosad

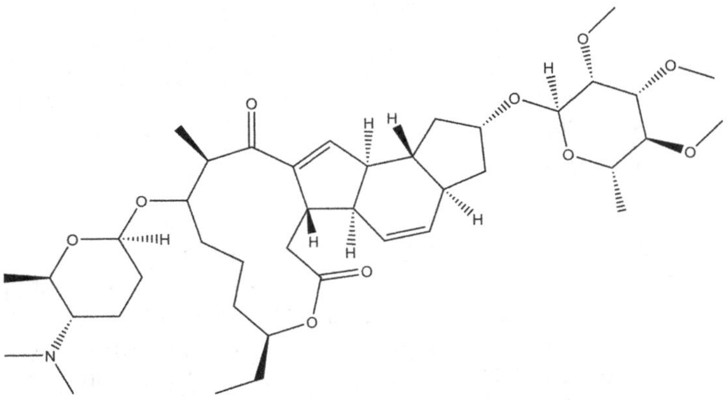

Spinosyn A (major component)

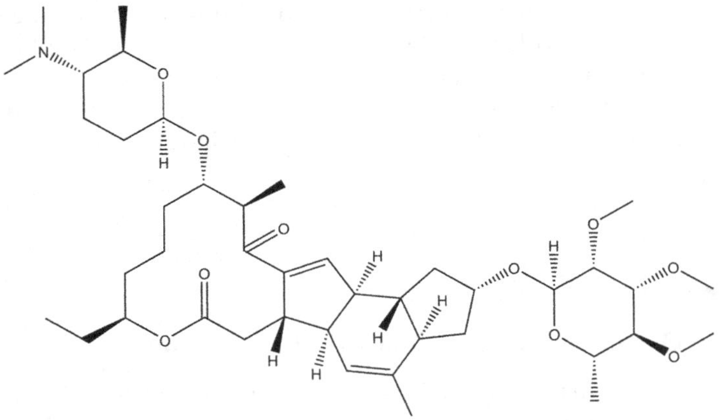

Spinosyn D (minor component)

**FIGURE 2.12**
Spinosyn major and minor molecular structures. (Credit to LHchem, CC BY-SA 4.0.)

**TABLE 2.1**

Physical and Chemical Properties of Spinosad

| Physical/Chemical Property | Spinosyn A Value | Spinosyn D Value |
|---|---|---|
| Melting point | 84°C–99.5°C | 161°C–170°C |
| Vapor pressure | $2.4 \times 10^{-10}$ mmHg at 25°C | $1.6 \times 10^{-10}$ mmHg at 25°C |
| Water solubility | 235 ppm at 25°C, pH 7 | 0.332 ppm at 25°C, pH 7 |
| Octanol-water partition coefficient ($K_{ow}$) | 54.6 | 90 |
| Henry's law constant ($K_h$) | $9.82 \times 10^{-10}$ atm·m$^3$/mol at 25°C, pH 7 | $4.87 \times 10^{-7}$ atm·m$^3$/mol at 25°C, pH 7 |
| Hydrolysis half-life | >30 days at 25°C, pH 7; 200 days at 25°C, pH 9 | >30 days at 25°C, pH 7; 259 days at 25°C, pH 9 |
| Photolysis half-life (aqueous) | 0.96 day | 0.84 day |
| Soil adsorption coefficient ($K_{oc}$) | 35,838 cm$^3$/g, averaged over different soil types | No data |
| Photolysis half-life (soil) | 8.68 days | 9.44 days |
| Aerobic soil metabolism half-life | 17.3 days in silt loam soil | 14.5 days in silt loam soil |
| Anaerobic soil metabolism half-life | 161 days | 250 days |

*Source:* Kollman, W. (2002). *Environmental Fate of Spinosad.* California DPR.

is effective against a broad range of insect species, including Diptera, Thysanoptera, Coleoptera, Orthoptera, Lepidoptera, and Hymenoptera. It operates by a neural mechanism, targeting the insect nervous system. The U.S. EPA has classified it as a reduced risk compound, as a naturally derived, low impact pesticide. The label carries the signal word "caution," which is the lowest human hazard word assigned by EPA. Chronic exposure studies on rats and mice have not shown inducement of tumors.

As a natural product, Spinosad is approved for use in organic agriculture by numerous nations. The properties of Spinosad translate to low persistence in the environment and low tendency to bioaccumulate (Table 2.1). It is also widely used to treat fleas on pets and head lice on humans. Spinosad was discovered in 1985 in isolates from crushed sugarcane, and it was first registered in 1996 in California for use on crops such as lettuce, broccoli, almonds, and oranges.

## 2.5 Basic Overview of Application of Pesticides

Common methods include spraying by ground rig or aircraft, drip irrigation, etc. (Crosby, 1998; Hartley and Graham-Bryce, 1980; Ware, 1983). Pesticide application involves spraying of a solution or a suspension in solvent, usually water, in a manner specified by the registrant on the product label. Or they may be formulated in a granular solid form that can be spread on the surface soil or foliage and "disked" into a specific depth. There are many variations, again spelled out on the label. The reader is referred to general books such as pesticide formulations by Hartley and Graham-Bryce (1980) and guides provided by the registrant and "best practices" provided by agricultural experiment stations and by the USDA Cooperative Extension Service. It is up to the landowner or applicator to know, understand, and follow best practices for formulation and application including warnings and restrictions on use. We will note in this book restrictions that might impact air contamination with pesticides and off-target contamination. A common example is restricted use of herbicide on windy days or too close to streams or other water sources.

Pesticide formulation affects its behavior in the environment. While the majority of agricultural pesticides are delivered by spraying or spreading on surfaces of soil or vegetation, there is a trend to applying them more in the irrigation system, e.g., by delivery to a drip or furrow or sprinkler irrigation system. This allows for delivery to the target efficiently with minimal off-target loss including to air. It works best for water-soluble chemicals that act systemically through uptake by the target crop or weed root zone.

## 2.6 Summary of Current Use of Pesticides

Although the landscape of pesticides is changing, earlier generations of pesticides remain in widespread use including sulfur, mineral oil, and copper sulfate (Table 2.2, Figure 2.13). Sulfur, horticultural oils, and fumigants were the most commonly applied pesticides in California for the year 2018 (California DPR).

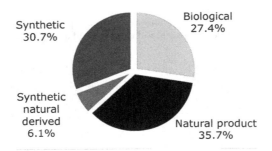

**FIGURE 2.13**
New active ingredient registrations for conventional pesticides and biopesticides from 1997 to 2010, organized by source. (Reprinted (adapted) with permission from Cantrell, C.L., Dayan, F.E., and Duke, S.O. (2012). Natural products as sources for new pesticides. *J. Nat. Prod.* 75, 1231–1242. Copyright (2012) American Chemical Society.)

**TABLE 2.2**

Top 25 Pesticides Used in California in 2018, Ranked by Pounds Applied

| Rank | Chemical | Pounds |
|---|---|---|
| 1 | Sulfur | 48,571,279 |
| 2 | Mineral oil | 29,611,582 |
| 3 | 1,3-Dichloropropene | 12,569,270 |
| 4 | Potassium n-methyldithiocarbamate | 8,527,736 |
| 5 | Petroleum oil, unclassified | 7,734,445 |
| 6 | Chloropicrin | 7,436,425 |
| 7 | Glyphosate, potassium salt | 6,109,802 |
| 8 | Glyphosate, isopropylamine salt | 5,766,081 |
| 9 | Calcium hydroxide | 5,016,453 |
| 10 | Metam-sodium | 3,765,705 |
| 11 | Kaolin | 3,268,360 |
| 12 | Sulfuryl fluoride | 2,991,914 |
| 13 | Alpha-(para-nonylphenyl)-omega-hydroxypoly(oxyethylene)[a] | 2,529,936 |
| 14 | Copper hydroxide | 2,346,020 |
| 15 | Pendimethalin | 2,244,649 |
| 16 | Methylated soybean oil[a] | 2,177,919 |
| 17 | Propanil | 1,829,002 |
| 18 | Copper sulfate (pentahydrate) | 1,801,464 |
| 19 | Lime-sulfur | 1,701,224 |
| 20 | Methyl bromide | 1,682,989 |
| 21 | Glufosinate-ammonium | 1,595,020 |
| 22 | Copper ethanolamine complexes, mixed | 1,510,703 |
| 23 | Petroleum distillates, refined | 1,395,969 |
| 24 | Mancozeb | 1,315,141 |
| 25 | Paraquat dichloride | 1,301,935 |

*Source:* CDPR (2018).

[a] Oils are carriers for active ingredients and also pesticides themselves. Hydrocarbon mixtures are also applied in agriculture as a component of dormant sprays on a variety of fruit and nut orchard trees to kill overwintering insects and diseases, and as weed oils.

# References

Boissoneault, L. (2018). *How the death of 6,000 sheep spurred the American debate on chemical weapons.* https://www.smithsonianmag.com/history/how-death-6000-sheep-spurred-american-debate-chemical-weapons-cold-war-180968717/.

Boffey, P.M. (1968). Nerve gas: Dugway accident linked to Utah sheep kill. *Science* 162, 1460–1464.

California Department of Pesticide Regulation (CDPR). (2018). The top 100 pesticides by pounds in total statewide pesticide use in 2018 https://www.cdpr.ca.gov/docs/pur/pur18rep/top_100_ais_lbs_2018.htm (Accessed February 3, 2021).

Cantrell, C.L., Dayan, F.E., and Duke, S.O. (2012). Natural products as sources for new pesticides. *J. Nat. Prod.* 75(6), 1231–1242. Doi: 10.1021/np300024u.

Carson, R. (1962). *Silent Spring.* New York: Houghton Mifflin.

CDPR. (2018). The top 100 pesticides by pounds in total statewide pesticide use in 2018. https://www.cdpr.ca.gov/docs/pur/pur18rep/18_pur.htm.

Crosby, D.G. (1998). *Environmental Toxicology and Chemistry.* Oxford and New York: Oxford University Press.

Darling, R. and Clendenin, B.(2020). *Overview of DPR's Neonicotinoid Reevaluation and Upcoming Mitigation Webinars* Pesticide Registration and Evaluation Committee. https://www.cdpr.ca.gov/docs/dept/prec/2020/071720_minutes.pdf

Hartley, G., and Graham-Bryce, I. (1980). *Physical Principles of Pesticide Behaviour.* London (UK): Academic Press Inc.

Kollman, W. (2002). *Environmental Fate of Spinosad.* Sacramento, CA: California Department of Pesticide Regulation.

McDonald-Gibson, C. (2013). 'Victory for bees' as European Union bans neonicotinoid pesticides blamed for destroying bee population. *The Independent.* 77 http://www.independent.co.uk/environment/nature/victory-for-bees-as-european-unionbans-neonicotinoid-pesticides-blamed-for-destroying-bee-8595408.html.

Seiber, J.N., Coats, J., Duke, S.O., and Gross, A.D. (2014). Biopesticides: State of the art and future opportunities. *J. Agri. Food Chem.* 62(48), 11613–11619. Doi: 10.1021/jf504252n.

Shibamoto, T., Yasuhara, A., and Katami, T. (2007). Dioxin formation from waste incineration. *Rev. Environ. Contam. Toxicol.* 190, 1-41.

United Nations Office of Disarmament Affairs. 2021. Chemical Weapons. https://www.un.org/disarmament/wmd/chemical/ (Accessed April 16, 2021).

US EPA. 2016. What are Biopesticides? https://www.epa.gov/ingredients-used-pesticide-products/what-are-biopesticides. (Accessed February 4, 2021).

Ware, G.W. (1983). *Pesticides, Theory and Application.* W.H. Freeman.

Wiener, S.W., and Hoffman, R.S. (2004). Nerve agents: A comprehensive review. *J. Intensive Care. Med.* 19(1), 22–37. Doi: 10.1177/0885066603258659.

# 3

## Physical and Chemical Properties of Pesticides and Other Contaminants: Volatilization, Adsorption, Environmental Distribution, and Reactivity

### 3.1 Introduction

When a chemical is released into the environment, it can contact and reside in one or more phases—air, water, soil, and biota. Chemicals can then be distributed in varying ratios into different compartments such as the atmosphere, hydrosphere, lithosphere, and biosphere (Figure 3.1).

Because many chemicals are released directly to the air first—by spraying, as with fumigants, or in combustion effluent—the air phase has been the focal point of much research over the years. The primary processes affecting distribution of these chemicals in the environment—volatilization, adsorption, and reactivity (stability)—are extensively discussed in this chapter. As we learn more, the details of chemical behavior in the environment emerge so that we can predict likely behavior and effects of contaminants before the release occurs.

### 3.2 Volatilization

Volatilization is a generalized process that is applied to both liquids and solids to describe the transition from the condensed phase to vapor. In closed containers, a substance will volatilize until equilibrium is reached. If the substance is exposed to the open atmosphere, however, equilibrium will not be achieved, since the open atmosphere acts as a transport medium and as an "infinite" reservoir for volatile chemicals. For liquids, volatilization is called evaporation; for solids, it is called sublimation. Common examples of volatile liquids include water and organic solvents. Depending on vapor

DOI: 10.1201/9781003217602-3

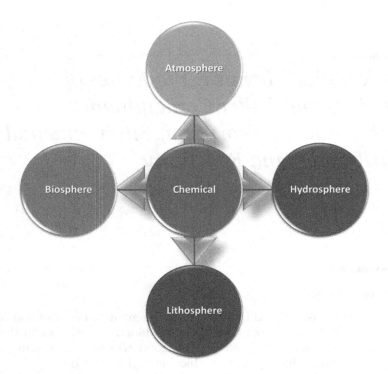

**FIGURE 3.1**
Four compartments of the environment that chemicals can partition into.

pressure and temperature, open containers of these liquids will, over time, reduce in volume and eventually disappear. Common examples of volatile solids include solid carbon dioxide (dry ice) and the common ingredient in moth balls—paradichlorobenzene. Depending on the temperature, dry ice can volatilize rather quickly because of its substantial vapor pressure (~5.7 × 10⁶ Pa [20°C]), whereas paradichlorobenzene, because of its much lower vapor pressure (~173 Pa [20°C]), volatilizes relatively slowly providing moth control over an extended period. Another common example includes agriculture, where applied pesticides can volatilize from treated soil, water, and foliage. Losses of pesticides by evaporation and sublimation can have economic and ecosystem/human health consequences. So, volatilization is a significant process in determining the fate of chemical compounds released in indoor and outdoor environments.

An estimate for losses by volatilization during and after application is approximately 80% of the total applied for some semivolatile organic compound (SVOC) pesticides. Toxaphene is an example (described below). Losses by this route are greatly influenced by such factors as physicochemical properties of active ingredients, temperature, wind speed, formulation, and type of surface (soil, water, vegetation, and others).

### 3.2.1 Volatilization of Pure Liquids and Solids

The term "pure" not only includes single compounds that are liquids or solids, depending on their properties, but it also encompasses liquids and solids composed of a mixture of like compounds. Examples of "pure" mixtures include hydrocarbon oils/fuels and paraffin. In other words, they are not dissolved in or blended with unlike materials. The main driving force for the volatilization of a pure substance is vapor pressure, which is a function of temperature. It is assumed, in this case, that the substance is volatilizing from a noninteractive (i.e., "inert") surface. This was observed in several field and laboratory studies with hydrocarbon mixtures and pure compounds (Mackay and Van Wesenbeeck, 2014; Seiber et al., 1986; Woodrow et al., 1997). The ideal gas law shows the relationship between vapor pressure (P) and temperature (T):

$$P = (n/V)RT \tag{3.1}$$

where R (8,314.6 L·Pa/K·mol) is the gas constant and n/V (moles/L) is the vapor density of the pure compound. If a pure chemical is sealed in a closed container at a particular temperature, volatilization will occur until equilibrium is reached (i.e., the number of molecules leaving the condensed phase is equal to the number returning to the condensed phase). Once equilibrium is reached, the vapor density can be determined, and the ideal gas law can be used to calculate the vapor pressure at the given temperature. Another relationship involving vapor pressure and temperature is the Clausius–Clapeyron equation:

$$\text{Ln} \left( P_2/P_1 \right) = \left( -\Delta H_{vap} / R \right) \left( [1/T_2] - [1/T_1] \right) \tag{3.2}$$

where $\Delta H_{vap}$ is the enthalpy of vaporization (cal/mole) and R is the gas constant (1.9872 cal/K·mole). If vapor pressure is known at one temperature, vapor pressure can be calculated at another temperature using this relationship.

In the above example of a closed container, if the vapor density is measured before equilibrium is reached, the flux (wt. lost/area/time) of the chemical can be determined using the following relationship (Rolston, 1986; Woodrow and Seiber, 1991):

$$F = (V/A)(\Delta c/\Delta t) \tag{3.3}$$

where F is flux, V is the vapor volume within the container, A is the surface area of the condensed phase, and $\Delta c/\Delta t$ is the time rate of change of the chemical concentration in the vapor (wt./volume/time). Another approach to the measurement of flux is the Knudsen effusion method, which involves the diffusion of a pure substance through a small orifice into the atmosphere:

$$Q = 1.98 \times 10^{-5} P \, (M/2\pi RT)^{1/2} \tag{3.4}$$

**TABLE 3.1**

Volatilization Rates of Chemical Residues on Noninteractive (Plant, Glass, Plastic) Surfaces

| | | Ln Flux ($\mu g/m^2 \cdot hr$) | |
|---|---|---|---|
| Compound | Surface | Observed | Knudsen[b] |
| Beacon oil[a] | Glass | 17.155 | 17.609 |
| Chevron oil[a] | Glass | 14.845 | 16.102 |
| Dodecane | Plastic | 13.805 | 15.297 |
| n-Octanol | Glass | 14.116 | 15.107 |
| Tridiphane | Giant foxtail | 8.872 | 9.159 |
| Trifluralin | Weedy turf | 7.371 | 8.487 |
| Pendimethalin | Turfgrass | 6.947 | 7.098 |
| 2,4-D (isooctyl) | Wheat | 6.507 | 6.799 |
| Diazinon | Dormant orchard | 6.812 | 6.155 |
| Toxaphene | Cotton | 5.293 | 5.278 |
| Dieldrin | Weedy turf | 5.147 | 5.447 |
| p,p'-DDT | Cotton | 3.824 | 2.708 |

*Source:* Reprinted (adapted) with permission from Woodrow, J.E., Seiber, J.N., and Baker, L.W. (1997). Correlation Techniques for Estimating Pesticide Volatilization Flux and Downwind Concentrations. *Environ. Sci. Technol.* 21(2), 523–529. Copyright (1997) American Chemical Society.
[a] Herbicidal mixtures of light hydrocarbons.
[b] Molecular effusion from pure material.

where Q is flux ($g/cm^2/sec$), P is the vapor pressure ($dyn/cm^2$) at T, M is the molecular weight, R is the gas constant ($8.314 \times 10^7$ erg/K·mol), and T is temperature. This equation describes molecular effusion from pure material and includes only intermolecular interactions in the material, as reflected in the vapor pressure. In an earlier study, the Knudsen equation was used to estimate flux for a series of chemicals applied to foliage and other noninteractive surfaces (Woodrow et al., 1997). The calculated values were compared to flux values measured in the laboratory and the field (Table 3.1).

Taking the antinatural logarithms ($e^x$), all of the calculated results compared to the observed flux values to within an order of magnitude, and for seven of the 12 compounds, the flux values compared to within a factor of 2. These results support the largely noninteractive nature of plant surfaces for freshly applied pesticides (less than 24 hours post application). This study led to a correlation of chemical vapor pressure with flux ($\mu g/m^2 \cdot hr$), measured in the laboratory and field for a series of liquid and solid pesticides:

$$y = 11.779 + 0.85543x, \; r^2 = 0.989 \qquad (3.5)$$

where y = Ln flux and x = Ln VP (Pa). This correlation can be used to estimate flux for new applications. Field measurements of pesticide flux were accomplished using techniques discussed in Section 3.2.4. Another useful relationship for flux is the one developed by Hartley:

$$\text{Flux}\left(\text{cmpd}\right) = \left[ P_{cmpd}\left(M_{cmpd}\right)^{1/2} / P_{water}\left(M_{water}\right)^{1/2} \right]\text{Flux}\left(\text{water}\right) \quad (3.6)$$

If the flux for a reference compound (e.g., water) is known at a given set of conditions, then, with molecular weights and vapor pressures, it is possible to calculate flux for the test compound. A recent study (Mackay and Van Wesenbeeck, 2014) correlated flux ($\mu g/m^2 \cdot hr$) with the product of vapor pressure and molecular weight (VP [Pa] $\times$ M [g/mol]):

$$\text{Flux} = 1{,}464\,(\text{VP} \times \text{M}) \quad (3.7)$$

This applies only to liquid surfaces that are unaffected by the underlying solid substrate, as occurs in the standard ASTM evaporation rate test and to quiescent liquid pools. The inclusion of M increased the slope of previous Ln flux vs. Ln VP regressions to a value close to 1.0. This correlation can be used for screening level assessment and ranking of liquid chemicals for evaporation rate, such as pesticides, fumigants, and hydrocarbon carrier fluids used in pesticide formulations, liquid consumer products used indoors, and accidental spills of liquids. In addition to vapor pressure, other factors that influence volatilization include movement of air over chemical deposits exposed to the open environment and the thickness of the deposit. The direct effect of wind flow rate on the volatilization of weed oil mixtures (e.g., Beacon oil, Chevron oil) was demonstrated in an earlier study (Woodrow et al., 1986). This study also showed that the weed oils (mixtures of hydrocarbons of varying molecular weight and vapor pressure) volatilized differentially from deposits on inert Teflon and glass surfaces (Figure 3.2). That is, components with higher vapor pressures and lower molecular weights volatilized early on, eventually leaving the original deposit enriched in the components of higher molecular weight and lower vapor pressure. This phenomenon was demonstrated both in the laboratory and in the field. In the field, weed oils applied to seed alfalfa led to significant volatilization, which is thought to have contributed to photochemical smog formation (see Table 2.2).

Another example of differential volatilization is the flux of hydrocarbons to the vapor above a residual jet fuel mixture in an airliner fuel tank, leading to concentrations of flammable volatile components in the headspace that could cause an explosion of considerable power (Woodrow, 2003). Figure 3.3 compares the chromatograms of jet fuel vapor and liquid. The liquid composition spans a carbon number range of about C5 to C16-C17, with the bulk of the hydrocarbon mass centered at about C12-C13 (hydrocarbon standards are shown superimposed on the gas chromatogram). By contrast, the vapor, in equilibrium with a pool of liquid fuel, spans the range C5 to about C11, with the bulk of the vapor below about C9.

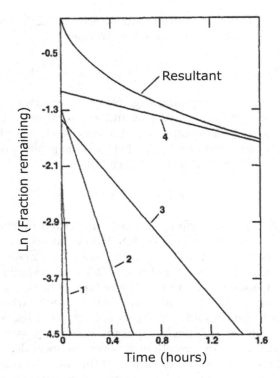

**FIGURE 3.2**
Volatilization of a weed oil mixture from a noninteractive surface. Numbers 1–4 represent fractional volatilization of the mixture, with 1 being the most volatile. (Reprinted (adapted) with permission from Woodrow, J.E., Seiber, J.N., and Kim, Y.-H. (1986). Measured and calculated evaporation losses of two petroleum hydrocarbon herbicide mixtures under laboratory and field conditions. *Environ. Sci. Technol.* 20(8), 783–789. Copyright (1986) American Chemical Society.)

A National Transportation Safety Board sponsored study showed that for a nearly empty center-wing 747 fuel tank (vapor volume to liquid volume [V/L] ratio was 274), the tank filled with an explosive fuel/air mix, due partially to the hot air conditioning units just below the tank and to the long wait on the JFK airport tarmac during hot summertime conditions. Even at a 14,000-foot altitude, these conditions essentially created a fuel–air explosive, offering an explanation for the tragic TWA Flight 800 accident over Long Island in 1996 that led to considerable loss of life. The source of ignition was unknown in this case, but found later to have resulted apparently from sparking of faulty electrical sensor wires in the tank itself.

An additional example of differential volatilization is toxaphene, a chlorinated hydrocarbon cotton insecticide that was heavily used in the 1950s–1980s, until it was banned in 1990. Toxaphene is comprised of 175-plus discrete polychlorinated terpenes—primarily camphenes and bornanes ($C_xH_yCl_z$). The

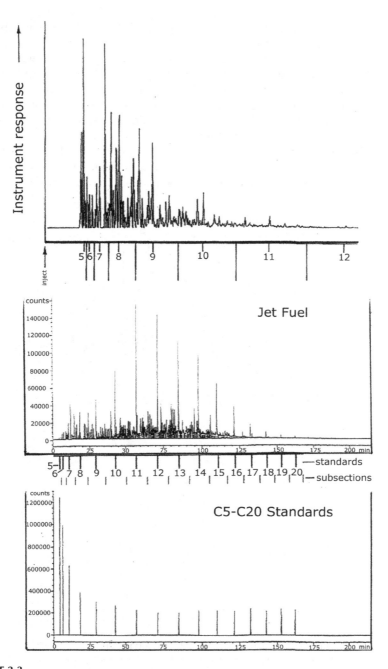

**FIGURE 3.3**
Gas chromatograms of jet fuel vapor (a) and of the liquid (b). (Reprinted (adapted) with permission from Woodrow, J.E. (2003). The laboratory characterization of jet fuel vapor and liquid. *Energy Fuels* 17(1), 216–224. Copyright (2003) American Chemical Society.)

initial residue after application shows a capillary GC fingerprint of technical toxaphene (Figure 3.4a). A sample of the air residue after weathering for 2 days or so shows a clear shift in residues favoring the more volatile, lower chlorinated congeners that have a shorter retention time, since the more volatile components are lost first from the residue by volatilization (Figure 3.4b).

Angermann et al. (2002) predicted that airborne residues of toxaphene could be transported significant distances from treated cotton fields in California. They provided evidence from findings of residues in pacific tree frogs taken from the Sierra Nevada Mountains, many miles downwind from the cotton fields. It has been postulated that deposition of anthropogenic compounds, such as from volatilized toxaphene, may be, at least in part, the cause of the

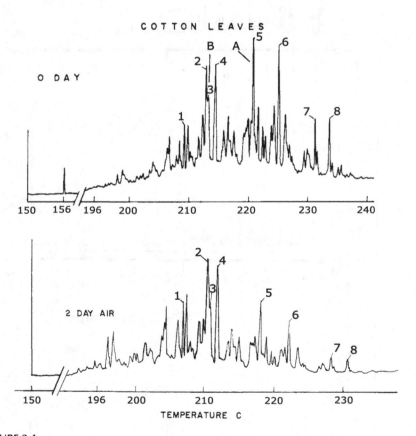

**FIGURE 3.4**
Capillary gas chromatograms of toxaphene residue in 0-day cotton leaf and 2-day air samples taken from within the same field. (Reprinted (adapted) with permission from Seiber, J.N., Madden, S.C., McChesney, M.M., and Winterlin, W.L. (1979) Toxaphene dissipation from treated cotton field environments: component residual behavior on leaves and in air, soil, and sediments determined by capillary gas chromatography. *J. Agric. Food Chem.* 27(2):284–291. Copyright (1979) American Chemical Society.)

decline of amphibians in the Sierras, due to the action of such compounds as endocrine system disruptors (see Chapter 4). These examples show the critical role in transport and fate played by volatility, as measured by vapor pressure, in the absence of other dissipation mechanisms such as chemical degradation, and when the surface from which evaporation is occurring is essentially noninteractive.

## 3.2.2 Volatilization from Water

### 3.2.2.1 Solutes

The fate of organic chemicals in an aqueous environment is determined by the complex interactions of chemical, biological, and physical processes. Chemical processes include hydrolysis, photolysis, chemical reactions, and the formation of metal ion complexes. Biological processes include microbial degradation and absorption by biota. Physical processes include dispersion, sorption by sediments, and volatilization.

For volatilization, vapor pressure is still the driving force that moves a compound into the vapor phase. But, an opposing role is played by the water's ability to attract or retard the chemical's tendency to evaporate. In this case the Henry's constant, H, is the controlling physicochemical property. Since H is proportional to P/S, high P (vapor pressure) can be counterbalanced by high S (water solubility). The interplay of vapor pressure and water solubility was shown in an earlier study that correlated flux normalized to water concentration (i.e., flux/[mg/L]) with vapor pressure/water sol. (VP/Sw) for a series of pesticides applied to water (Woodrow et al., 1997). It was assumed that a chemical's vapor pressure would be modified primarily by water solubility. That is, the lower the residue concentration in water, the lower the effective vapor pressure (through mole fraction and activity), and thus the lower the volatilization rate. This is illustrated by the form of Raoult's law for nonideal solutions:

$$P_i = x_i \gamma_i P^o \tag{3.8}$$

where $P_i$ is the partial pressure of component i in the mixture, $x_i$ is its mole fraction, $\gamma_i$ is its solution activity coefficient, and $P^o$ is its saturation vapor pressure. For water-compatible compounds (i.e., exothermic mixing), the activity coefficient is commonly less than unity, showing the effect of solubilization (van der Waal's type interactions and hydrogen bonding). For non-water compatible compounds of low solubility (i.e., endothermic mixing), such as alkanes and chlorinated alkanes, the activity coefficient can be much greater than unity, leading to higher than expected values for $P_i$ (positive deviation from the ideal).

The Woodrow et al. (1997) study led to very good correlation for about 12 cases, but only when the measured flux was normalized to actual water concentration (Table 3.2):

**TABLE 3.2**

Volatilization Rates for Pesticides Applied to Water

| Compounds | Ln (VP/Sw) | Absolute Flux (µg/m²·hr) | Water Conc. (mg/L) | Ln (flux/[mg/L]) |
|---|---|---|---|---|
| Deltamethrin | −6.91 | 1.3–3.5 | 0.00034–0.00095 | 8.23 |
| Deltamethrin | −6.91 | 1.3–6.4 | 0.00034–0.00173 | 8.23 |
| Diazinon | −8.94 | $4.8 \times 10^3$ | 13.5 | 5.87 |
| Eptam | −4.42 | $2.6 \times 10^4$ | 2.17 | 9.39 |
| Ethyl-parathion | −9.98 | 605 | 3.5 | 5.15 |
| Me-parathion | −10.3 | $1.12 \times 10^3$ | 14.5 | 4.35 |
| Mevinphos | −14.53 | 399 | 174 | 0.83 |
| Molinate | −6.83 | $5.75 \times 10^3$ | 3.43 | 7.42 |
| Molinate | −6.83 | $3.93 \times 10^3$ | 1.80 | 7.69 |
| Molinate | −7.57 | $1.58 \times 10^4$ | 25.1 | 6.44 |
| Thiobencarb | −9.62 | 160 | 0.57 | 5.64 |
| Thiobencarb | −9.62 | $3.88 \times 10^3$ | 16.3 | 5.47 |

*Source:* Reprinted (adapted) with permission from Woodrow, J.E., Seiber, J.N., and Baker, L.W. (1997). Correlation Techniques for Estimating Pesticide Volatilization Flux and Downwind Concentrations. *Environ. Sci. Technol.* 21(2), 523–529. Copyright (1997) American Chemical Society.

$$y = 13.643 + 0.8687x, \ r^2 = 0.970 \tag{3.9}$$

where $y = \mathrm{Ln}$ (flux/[mg/L]) and $x = \mathrm{Ln}$ (VP/Sw). This correlation was especially encouraging in view of the fact that the database contained a mix of laboratory and field volatilization measurements. While a good correlation of the data required normalizing flux to water concentration, it is the absolute flux values (µg/m²·hr) that give a true measure of the mass movement of chemical from water to air. For example, since the water solubility of deltamethrin is so exceedingly low (Sw = 0.002 mg/L), the concentration in water will also be low in field situations, and the absolute flux of this compound would not be expected to be very high on just a mass/area basis, even though normalized flux placed deltamethrin near eptam (Sw = 375 mg/L). This is borne out by the absolute flux values measured in the field for deltamethrin (1.3–6.4 µg/m²·hr) and eptam (2.6 × 10⁴ µg/m²·hr). The correlation should be used to derive absolute flux values from chemical properties and measured water concentrations. Field measurements of flux were accomplished using techniques discussed in Section 3.2.4.

So, the volatilization flux of a water-borne chemical depends on both vapor pressure and water solubility. A nonpesticide example is the Deepwater Horizon oil rig blowout in April 2010 that resulted in a 5-month outflow of crude oil from the ocean floor before it was capped. The lighter hydrocarbons in the water column volatilized rapidly after the spill because of their relatively high Henry's constants and their nonideal aqueous solutions. This phenomenon was also observed during the Exxon Valdez oil spill in Prince William Sound, Alaska, in 1989. In addition to volatilization from the ocean water column, significant volatilization also occurred from the crude oil deposited on the beaches. Workers hired to clean the beaches experienced long-lasting health effects due to their exposure to the volatile components of the crude, even though they wore protective equipment (Ott, 2005).

This phenomenon is further illustrated by the volatilization of dimethylmercury from ocean water and the deposition of the mono-methyl cation in rainwater and fogwater along the California coast (see Chapter 7). Dimethylmercury has a high Henry's constant driving the flux to the air as ocean water hits the coastline. This explains in part the enrichment of mercury along the Pacific Ocean coastline and in air along the coast. Volatilization of organic compounds from water is usually described by the two-film model (Lewis and Whitman, 1924), which assumes that all of the resistance to volatilization is in thin films of water and air at the interface between the phases. The resistances of these films are assumed to be in series and, therefore, additive. Liquid- and gas-phase mass transfer coefficients are related to chemical properties and fluid flow conditions:

$$k_l = D_l/a_l, \quad k_g = D_g/a_g \tag{3.10}$$

The mass transfer coefficients are $k_l$ and $k_g$ for liquid phase and gas phase, respectively (cm/sec), $D_l$ and $D_g$ are the corresponding diffusion coefficients (cm$^2$/sec), and $a_l$ and $a_g$ are the hypothetical thicknesses of the liquid film and gas film, respectively, adjacent to the interface (cm) (Figure 3.5).

In addition to the two films, other factors that affect volatilization of organic compounds from water are water temperature and wind speed, leading to turbulence in air and water. The hypothetical film thicknesses are assumed to be a function of the extent of turbulent kinetic energy and subsequent

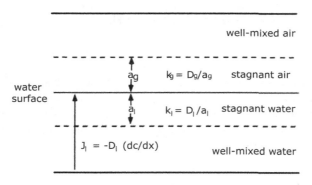

**FIGURE 3.5**
Schematic representation of the two-film model for the flux of organic compounds from water.

mixing on either side of the interface, decreasing in width with an increase in turbulent kinetic energy. For dilute aqueous solutions, the film thicknesses are assumed to be independent of chemical concentrations. In the absence of turbulence, pure diffusion, as measured by liquid diffusion coefficients ($D_l$), is a dominant process in volatilization. Since $D_g > D_l$, the rate-limiting step for flux from an aqueous solution will be diffusion from the liquid:

$$J_l = - D_l(dc/dx) \tag{3.11}$$

for an ideal solution, where $J_l$ is diffusive flux from water and dc/dx is the liquid concentration gradient. For a nonideal solution, the relationship is

$$J_i = - (Dc_i/RT)(\delta\mu_i/\delta x) \tag{3.12}$$

where R and T are the gas constant and temperature, respectively, and $\mu_i$ is the chemical potential for the ith species. The chemical potential $\mu_i$ can be written as

$$\mu_i = \mu_i^\circ + RT \ln \alpha_i \tag{3.13}$$

where activity $\alpha_i = X_i\gamma_i$, the product of mole fraction and activity coefficient.

This all comes together in a foggy atmosphere or cloud, in which suspended water droplets both accumulate chemicals from the air, and also release the accumulated chemicals to the air. Glotfelty et al. (1987) measured the concentration of a number of pesticides, polychlorinated biphenyls (PCBs) and other semivolatile chemicals in fogwater, and found that fogwater accumulated some of the contaminants to a greater degree than predicted by Henry's law, apparently reflecting operation of the water surface film as a third phase in the water-to-air partition coefficient (see Chapter 7). During rainfall, rain drops can absorb atmospheric components in a manner that can be predicted by the components' Henry's law constants, resulting in a parameter known as washout ratio (W), which is a function of H:

$$W \propto 1/H \tag{3.14}$$

As with many environmental distribution coefficients, these processes occur as described above for pure water and pure air systems. If the water contains dissolved solutes such as salt, or surface-active detergent molecules, or suspended clay or other soil components, the observed air/water distribution can be significantly affected.

### 3.2.2.2 Azeotropes

If the total vapor pressure of aqueous mixtures deviates positively or negatively from Raoult's law, azeotropes can be formed at a particular mixture composition. When boiled, an azeotropic mixture will produce vapor of the same composition as the liquid mixture, and this situation will not change with further evaporation. For a hypothetical binary mixture of components, A and B, total vapor pressure ($P_t$) can be described as follows:

$$P_t = X_A \gamma_A P_A + X_B \gamma_B P_B \qquad (3.15)$$

where $P_A$ and $P_B$ are the saturation vapor pressures of components A and B, respectively; the other terms are as described above. If one plots mixture composition vs. total pressure, the result will look something like Figure 3.6.

The center line is the ideal. The two curves show positive and negative deviations from the ideal. A positive deviation will show a maximum pressure for the mixture and a negative deviation will show a minimum pressure. Horizontal tangents at the maximum and at the minimum will give the azeotropic mixture composition at those points.

In phase diagrams, the tangent points are where vapor composition is the same as the liquid composition. For example, Figure 3.7 is a phase diagram for a negative azeotrope (i.e., maximum temperature, minimum pressure). The boiling point for a positive azeotrope will be less than the boiling points of the components; the boiling point of a negative azeotrope will be greater than the component boiling points. If the plot in Figure 3.7 is done at another fixed temperature, the total vapor pressure of the mixture will change, and it is possible that the composition at which the azeotrope occurs will also change. Azeotropes can be formed by mixtures of three or more components, but the mixture behavior becomes complex.

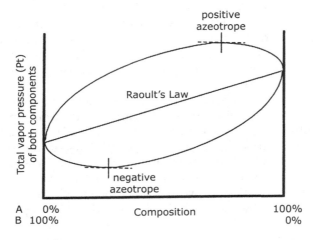

**FIGURE 3.6**
Composition vs. total vapor pressure for binary mixtures of components A and B at constant temperature.

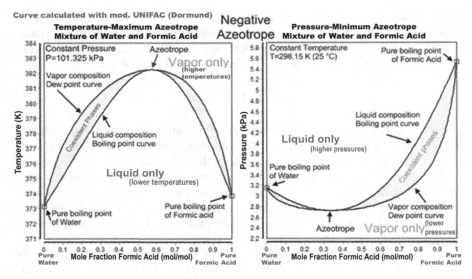

**FIGURE 3.7**
Vapor–liquid equilibrium of formic acid in water.

### 3.2.3 Volatilization from Soil

In general, volatilization from soil increases with soil moisture, as the moisture is able to displace a chemical sorbed to the soil surface. Chemical residues are also mobilized upward to the air/soil surface interface by mass transport with the evaporating water—a process referred to as wicking. Wind movement over the surface increases volatilization, while depth of incorporation of the chemical retards volatilization. Temperature generally increases volatilization, while atmospheric conditions (inversions or stable atmospheres) have a variable effect. Generally, any factor that promotes evaporation of water from soil will also promote contaminant volatilization to the surrounding air as well.

A dramatic example of chemical volatilization caused by soil moisture was shown in an earlier study of the volatilization of trifluralin preemergence herbicide applied to fallow soil (Soderquist et al., 1975). Two soil plots were treated—one was surface-only and the other was disced in to about a 15 cm depth (soil-incorporated) after application. When water was applied to the two plots—the soil-incorporated plot experienced a rain event—concentrations of trifluralin at 0.5 m above the plots increased substantially. Vapor concentrations above the surface-only plot exceeded 2,500 ng/m³, while concentrations above the soil-incorporated plot were above about 150 ng/m³, showing the less dramatic effect of water on disced-in material.

Two other studies were made of the behavior of pesticides applied to soil (Woodrow et al., 1997, 2011). The physicochemical properties of the pesticides were correlated with measured emission rates ($\mu g/m^2 \cdot hr$) to develop correlations that could be used to estimate emissions for new application situations. The main driving force for volatilization from soil is vapor pressure, but volatilization will be lessened by the soil adsorption coefficient (Koc [solubility in soil organic carbon]), water solubility (Sw), and depth of incorporation (d). For semivolatile pesticides, two relationships were derived (Woodrow et al., 1997):

$$\text{Ln flux} = 28.355 + 1.6158 \text{ Ln } R_{surf}, \quad r^2 = 0.988 \tag{3.16}$$

for surface-applied chemical, where $R_{surf} = VP/(Koc \times Sw)$, and

$$\text{Ln flux} = 19.35 + 1.0533 \text{ Ln } R_{inc}, \quad r^2 = 0.93 \tag{3.17}$$

for soil-incorporated chemical, where $R_{inc} = (VP \times AR)/(Koc \times Sw \times d)$. VP is vapor pressure (Pa), Koc is soil adsorption coefficient (mL/g), Sw is water solubility (mg/L), AR is application rate (kg/ha), and d is depth of incorporation (cm). A similar relationship was derived for a series of soil fumigants applied by injection and surface chemigation (Woodrow et al., 2011):

$$\text{Ln flux} = 3.598 + 0.940 \text{ Ln } R, \quad r^2 = 0.994 \tag{3.18}$$

where flux has the units $\mu g/m^2 \cdot sec$ and $R = (VP \times AR)/(Koc \times Sw \times d)$. In all cases, field measurement of flux was accomplished using techniques discussed below.

### 3.2.4 Volatilization Flux Methods

Since volatilization is a significant process in determining the fate of chemical compounds in the environment, several methods have been developed for determining flux of chemicals, especially for pesticides applied to soil (see Chapters 6 and 8). One simple method consists of laboratory soil columns treated with the chemicals of interest. The top of the columns can be sealed with a chamber through which air is passed, transporting volatilized chemical vapors into a trap for analysis. If the surface area (A, $cm^2$) of the top of the soil column is known, along with air flow rate (Q, $cm^3/min$) over the surface of the soil and chemical vapor concentration (C, $\mu g/cm^3$) in the exhaust air, flux (F, $\mu g/cm^2/min$) can be calculated:

$$F = QC/A \tag{3.19}$$

Soil type, temperature, airflow rate, soil moisture content, chemical application rate, etc., can be varied to model field environments. A similar field method is the use of closed chambers, with chemical vapor sampling ports, placed over portions of treated fields (Woodrow and Seiber, 1991). Depending on the volatility of the chemical, the outside air intake port may need to be filtered to avoid biasing flux determinations within the chamber.

However, for open field measurements of flux (soil, plant canopy), different approaches are required. While several different methods have been developed for chemical flux determination in the open field, three that are commonly used are the aerodynamic gradient (AG), theoretical profile shape (TPS), and integrated horizontal flux (IHF) methods. These micrometeorological methods give essentially equivalent results (Majewski et al., 1990). Choosing one method over another depends on preference, field configuration, availability of equipment, and budgetary considerations.

AG chemical fluxes ($F_{AG}$) can be estimated using a modified form of the Thornthwaite–Holzman equation (Thornthwaite and Holzman, 1939) corrected for atmospheric stability conditions (Majewski et al., 1995):

$$F_{AG} = \left(k^2 \Delta c \, \Delta u\right) / \left(\varphi_m \varphi_p \left[ \ln \left(z_2/z_1\right) \right]^2\right) \tag{3.20}$$

where k is the von Kármán constant (~0.41), $\Delta c$ and $\Delta u$ are the average chemical concentration and wind speed differences, respectively, between heights $z_1$ and $z_2$ above the treated surface, and $\varphi_m$ and $\varphi_p$ are atmospheric stability functions. To use the AG method, it is necessary to determine vertical gradients of wind speed and chemical concentrations. Also, a fetch of about 100:1 needs to be established. The fetch is the ratio of the upwind edge distance of the field to the height of the sampling tower, usually located at the center of the field (Figure 3.8).

In an earlier study, the AG method was used to estimate the post-application methyl bromide (MeBr) volatilization loss rates from two different application

**FIGURE 3.8**
AG equipment setup for monitoring the volatilization of MeBr from a treated and tarped field. Michael Majewski is standing near right center of photo.

practices (Majewski et al., 1995). As the fumigant was being injected at 25–30 cm depth, one field was covered simultaneously with a high-barrier plastic film tarp (Figure 3.8). Another field was fumigated in like manner and left uncovered, but the furrows made by the injection shanks were bedded over. The cumulative volatilization losses from the tarped field were 22% of the nominal application within the first 5 days of the experiment. In contrast, the nontarped field lost 89% of the nominal application by volatilization in 5 days. These results emphasize the importance of "sealing" a field with a tarp to reduce losses of a volatile fumigant, such as MeBr, which is a known ozone-depleting chemical (see Chapter 8 for more on MeBr).

The TPS method (Majewski et al., 1982; Yates, 2006) requires a circular field and wind speed and chemical vapor concentration are measured at one height ($Z_{inst}$) above the field:

$$F_z = (u\ c)/R_{inst} \tag{3.21}$$

where u ($Z_{inst}$) is the average wind speed, c ($Z_{inst}$) is the average chemical vapor concentration, and $R_{inst}$ is the ratio of horizontal to vertical flux at the measurement height ($Z_{inst}$). $R_{inst}$ depends on the surface roughness and the radius of the circular field but does not depend on the wind speed. The roughness height is derived from the wind speed profile measurements made prior to the field experiment. This method was used in one study to determine the volatilization losses of the herbicide triallate applied to the surface of fallow soil (Yates, 2006). This study also compared the TPS method with the IHF method and found that the two methods gave similar results. A separate study compared the TPS method with the AG method and found that, while the two methods gave equivalent results, the TPS method was easier and less costly to use (Majewski et al., 1989).

For the IHF method (Majewski et al., 1990; Woodrow et al., 1997), a chemical vapor sampling mast is placed at the downwind edge of the source, and vapor concentrations are measured at various heights. Wind speed is also measured at the same sampling heights. At the downwind edge of the source, the chemical vapor plume height will be about 10% of the depth of the source, depending on atmospheric stability. Using the plume height at the downwind edge and the concentration and wind speed profiles, it is possible to calculate the IHF ($F_z$):

$$F_z = 1/X \int c_i \times u_i\, \partial z \tag{3.22}$$

where $c_i$ and $u_i$ are average concentration and wind speed, respectively, at height z. X is the distance to the upwind edge of the source, and the product $c_i \times u_i$ is integrated over the plume height, z. The setup and equipment requirement for the IHF method is essentially the same as that for the AG method, except that IHF does not require the 100:1 fetch that the AG method needs to give reliable results.

The focus of this chapter is chemical volatilization. It is one of a number of possible processes (physical, chemical, biological) that contribute to the fate of a chemical. The diverse examples showed that volatilization is a general process for both liquids and solids. Volatilization can explain the initial occurrence of chemicals in the atmosphere, where they eventually will be dispersed and possibly deposited at some remote location, depending on stability. The banned insecticide DDT is an example of this. Residues of DDT, and of its major conversion product DDE, have been found as far away as Antarctica. Another example is the ozone-depleting soil fumigant MeBr. After volatilizing from treated soil, MeBr has enough stability in the environment to make its way to the stratosphere. These examples, and many others (e.g., treated agricultural fields, pavement coatings, and spills), have prompted the development of methods for determining volatilization flux— flux chambers, micrometeorological methods (AG, TPS, IHF), numeric and computer modeling, and others.

## 3.3 Adsorption/Absorption

Adsorption can be either physical or chemical in nature. In the bulk of the adsorbent, other atoms fill the bonding requirements (ionic, covalent, metallic) of the constituent atoms. However, atoms on the surface of the adsorbent are not wholly surrounded by other adsorbent atoms and therefore can attract adsorbates. For the condensation of gases to liquids, for example, physical adsorption depends on the van der Waals force of attraction between the solid adsorbent and the adsorbate molecules. There is no chemical specificity in physical adsorption: any chemical contaminants in air tend to be adsorbed on any solid if the temperature is sufficiently low or the pressure of the contaminant sufficiently high. In chemical adsorption, chemical contaminants in air or water are held to a solid surface by chemical forces (ionic, covalent) that are specific for each surface and each chemical. Chemical adsorption occurs usually at higher temperatures than those at which physical adsorption occurs; furthermore, chemical adsorption is ordinarily a slower process than physical adsorption and, like most chemical reactions, may involve an energy of activation.

Adsorption differs from absorption, in which a fluid (the *absorbate*) is dissolved by or permeates a liquid or solid (the *absorbent*). Absorption is made between two phases of matter: for example, a liquid absorbs a vapor, or a solid absorbs a liquid. When a liquid solvent absorbs a chemical vapor mixture or part of it, a mass of chemical moves into the liquid. For example, water may absorb oxygen from the air. This mass transfer takes place at the interface between the liquid and the chemical vapor, at a rate depending on both the chemical and the liquid. This type of absorption depends on the solubility of the chemicals and their pressure and temperature. The rate and amount of absorption also depend on the surface area of the interface and its duration in time. For example, when the water is finely divided and mixed with air, as may happen in a waterfall or strong ocean surf, the water absorbs more oxygen.

When a solid absorbs a liquid mixture or part of it (e.g., a chemical contaminant dissolved in the liquid), a mass of liquid moves into the solid. For example, a clay pot used to store water may absorb some of the water. This mass transfer takes place at the interface between the solid and the liquid, at a rate depending on both the solid and the liquid. For example, pots made from certain clays are more absorbent than others. Absorption involves the whole volume of the material, while adsorption is a surface-based process. The term *sorption* encompasses both processes, while *desorption* is the reverse of them.

## 3.4 Environmental Distribution

In the environment, sorption of contaminants by various media involves equilibria. The important distributions affecting a contaminant released to the environment are summarized in Figure 3.9. A contaminant released to the environment in water effluent will initially favor the water environment, but with time will undergo distributions as shown in Figure 3.9. Some will volatilize to the air governed by the Henry's law constant, some will sorb to soil and sediment governed by the sorption coefficient, and some will partition into the biota governed by the bioconcentration factor (BCF). Distribution from air to soil is principally governed by vapor pressure, and biota to soil by a BCF appropriate to soil biota. Accompanying these distributions are other

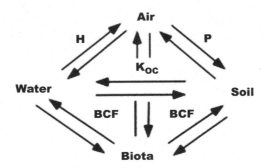

**FIGURE 3.9**
The important distributions affecting a contaminant released to the environment. (Abbreviations: P, vapor pressure; H, Henry's constant; $K_{oc}$, soil sorption coefficient on organic fraction basis; BCF, bioconcentration factor.)

fate processes, including chemical/photochemical or microbial degradation. Such a distribution is similar to what is observed for chemical contaminants in a laboratory microcosm consisting of sand/soil, plants, water, and animals (see Chapter 5).

### 3.4.1 Soil

#### 3.4.1.1 Distribution between Soil/Sediment and Water

It is important to be able to measure the sorption equilibrium for a chemical contaminant between water and soil/sediment. However, competing with this process is the possible loss of contaminant from water to the air. This is determined by the contaminant's air–water partition coefficient ($K_{aw}$), which can be expressed as

$$K_{aw} = C_a/C_w = H/RT \tag{3.23}$$

where $C_a$ is concentration in air, $C_w$ is concentration in water, H (Pa·m$^3$/mol) is the Henry's law coefficient, and R is the gas constant (8.314 Pa·m$^3$/mol K).

The sorption equilibrium for a contaminant between water and soil can be illustrated as follows:

$$\text{Solid (adsorbent)} + \text{solute (free)} \overset{K_f}{\leftrightarrow} \text{Solid-solute (bound or sorbed)} \tag{3.24}$$

Typically this is done by preparing multiple replicates of adsorbent (e.g., soil) in an Erlenmeyer flask or other suitable container, then adding six or more different initial concentrations of solute in water, containing a known concentration of solute stated as μg/mL. The mixtures are then shaken for several hours to establish equilibration between water and soil, then filtered or decanted to yield the aqueous supernatant. The concentration of solute is then measured as C (μg/mL) in water, and concentration in soil as x/m in μg/g. Soil concentration (x/m) is then plotted vs. C (water concentration). Typically this gives a plot similar to that shown in Figure 3.10, with departure

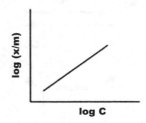

**FIGURE 3.10**
Plot of log (x/m) vs. log C.

from linearity commonly observed as adsorptive sites on soil become saturated with solute:

$$x/m = K_f C^{1/n} \tag{3.25}$$

where $K_f$ is the Freundlich adsorption isotherm constant. While other adsorption isotherms are theoretically based, the Freundlich isotherm is empirically determined. This may be written in log form for calculations:

$$\log (x/m) = \log K_f + 1/n \log C \tag{3.26}$$

The plot of log (x/m) vs. log C has a slope of 1/n (n, which has no physical meaning, varies between 0.7 and 1.1) and an intercept equal to log $K_f$. $K_f$ varies from 0.1 (weak adsorption) to 1,000 or more. Contaminants with high water solubility (S) generally give low $K_f$, and contaminants with low S generally give high $K_f$ (Table 3.3).

$K_f$ is pH dependent for ionizable solutes, such as phenol whose water solubility increases as pH increases, and the phenolate anion is the dominant form in solution. If we plot log $K_f$ vs. log S for different chemicals, the relationship typically appears as shown in Figure 3.11.

For a given organic chemical, $K_f$ can vary considerably from soil to soil or sediment to sediment, depending on the properties of the sorbent. Also, the amount of organic matter can vary widely from one type of soil to another (e.g., <0.1% in clay to 20% in muck). For many organic chemicals, and in particular

**TABLE 3.3**

Examples of the Sorption Equilibrium for a Chemical
Contaminant between Water and Soil/Sediment

| Chemical | $K_f$ | S (ppm) |
|---|---|---|
| Phenol | 0.2 | 82,000 |
| DDT | 100,000 | 0.0017 |

Abbreviations: $K_f$, Freundlich adsorption isotherm constant; S, water solubility in parts per million (ppm).

**FIGURE 3.11**
Plot of log $K_f$ vs. log S for different chemicals.

neutral hydrophobic organics (e.g., DDT), sorption is directly proportional to the quantity of organic matter associated with the solid. Normalizing the soil or sediment-specific sorption coefficient to the organic carbon (oc) content of the sorbent yields a new coefficient, $K_{oc}$, that is considered to be a unique property of the organic chemical being sorbed:

$$K_{oc} = K_f / oc \text{ or } K_d / oc \tag{3.27}$$

$$K_f \text{ from: } x/m = K_f C^{1/n} \tag{3.28}$$

$$K_d \text{ from: } K_d = C_s / C_w \tag{3.29}$$

where $C_s$ and $C_w$ are concentrations in the sorbed phase and in the aqueous phase, respectively.

$K_{oc}$ is the organic carbon normalized sorption coefficient, $K_d$ and $K_f$ are the linear and Freundlich (nonlinear) sorption coefficients specific to a particular sorbent and chemical combination, and oc (determined by weight loss during combustion of the soil) is the organic carbon content of that sorbent in units of g oc/g dry soil. Since $K_d$ is independent of concentration, it should be used in order for $K_{oc}$ to be a true constant. But, $K_f$ can also be used if $1/n$ is known (i.e., $x/m = K_f C^{1/n}$).

The most widely used approach for estimating $K_{oc}$ is based on correlations with physicochemical properties, such as $K_{ow}$ (octanol–water partition coefficient) and S (water solubility). A number of regression models have been derived relating $K_{ow}$ and S to $K_{oc}$. Some models can be generally applied and others are specific to chemical classes. Compilations of regression models for $K_{oc}$ can be found (Gawlik et al., 1997; Lyman, 1985; Mackay and Boethling, 2000). An extensive list of experimental log $K_{oc}$ values for a variety of organic chemicals, including some common pesticides, can be found in Mackay and Boethling (2000) and literature citations therein.

### 3.4.1.2 *Distribution between Soil/Sediment and Groundwater*

With chemical contaminants in soil (e.g., pesticides and chemical spills), there is concern over possible groundwater contamination. Chemical properties that favor groundwater contamination include low $K_{oc}$, high water solubility (S), moderate vapor pressure (P), and recalcitrance to microbial and chemical decomposition. Together, these properties create the "perfect storm" for groundwater contamination. However, in reality, most environmentally persistent chemicals may have one property that dominates its behavior regarding possible groundwater contamination. DDT, for example, has a low vapor pressure (~0.025 mPa) and low water solubility (0.0017 mg/L), but its $K_{oc}$ value is high (~240,000; log $K_{oc}$=5.38). Primarily because of its high $K_{oc}$ value, it is highly unlikely that DDT will be found in groundwater. Aldicarb, by comparison, has a low to moderate $K_{oc}$ value (30; log $K_{oc}$=1.48) and moderate vapor pressure (13 mPa), but its water solubility is comparatively high (~6 g/L). Primarily because of this, measurable residues of aldicarb have been found in groundwater. The soil fumigants ethylene dibromide (EDB) and MeBr have comparable $K_{oc}$ values (44 and 56, respectively) and similar water solubilities (4 and 17 g/L, respectively). But, the vapor pressure of MeBr (~200 kPa) is over 100 times greater than that for EDB (~1.56 kPa). Measurable residues of EDB have been found in groundwater, but not MeBr.

### 3.4.1.3 Distribution between Soil/Sediment and Vapor

Sorption of chemical vapor by soil can be described by the Freundlich isotherm equation, which takes the form

$$x/m = K^*P^{1/n} \tag{3.30}$$

or

$$\log (x/m) = \log K^* + 1/n \log P \tag{3.31}$$

where P is the chemical's partial pressure. This isotherm adequately describes adsorption at low P, typical of chemical contaminants in an air basin or airborne pesticide residues near treated fields. Chemicals in air can be transported to the soil surface by diffusion, by a process following Fick's first law, and advection (mass transport). Diffusion is defined as the movement of chemical along a concentration gradient from higher to lower concentration. Movement to the soil is measured as flux, where flux is governed by the chemical's diffusion coefficient (D [cm²/sec]) and the concentration gradient (dc/dx):

$$\text{Flux} = -D \, dc/dx \tag{3.32}$$

where flux has the units mass/soil area/time. If the chemical in air has a moderate vapor pressure and a relatively high $K_{oc}$ value, the soil will become a sink for the chemical creating a steep concentration gradient. Flux can occur out of the soil as well. Examples include not only gaseous products of microbial metabolism (e.g., $NO_x$ and $H_2S$) but also the soil fumigants that are used in agriculture to control soil pests. Fumigants are characterized by high vapor pressures, high diffusion coefficients, and relatively low $K_{oc}$ values. These properties allow fumigants to permeate soil for pest control, but they also drive the fumigants along a steep concentration gradient to the soil surface, where they move away by diffusion and advection.

### 3.4.1.4 Distribution between Soil/Sediment and Fog/Rainwater

This is a special case for adsorption/absorption equilibria involving all of the above—water, vapor, and suspended particulates (i.e., soil and dust). An example of how fog can accumulate and transport chemicals was in a recent study that showed the occurrence of mono-methyl mercury cation in marine fog along the California coast. It originated from dimethyl mercury that was formed in the marine environment (see Chapter 7). In another earlier study of pesticide distribution between vapor and fogwater, it was found that enrichment of some chemicals in the water phase occurred over what would be expected from just a simple water–air distribution (i.e., $K_{aw}$) (Glotfelty et al., 1987) (see Chapter 7). It turned out that a significant particulate fraction in the water operated as a third phase, which affected the denominator in the equation

$$K_{aw} = C_a / C_w \qquad (3.33)$$

The third phase due to water-borne particulates may be significant in the calculations of air–water distributions.

Fogs that occur in agricultural areas will absorb volatilized pesticides and transport them from their intended targets. Heavy tule fogs are a common occurrence in the heavily agricultural Central Valley of California during the winter months. Before people realized that fog could be a transport medium for chemicals, some farmers were fined by the state for misusing pesticides. It turned out that the pesticide residues found on crops for which the chemical was not registered were deposited by condensing fog from nearby fields that had been legally treated.

Rainwater is another aqueous medium that can scrub/wash the air of particulates and chemicals during a rainfall. An example is a study that showed the occurrence of water-soluble trifluoroacetic acid (TFA) in vernal pools formed by rain events downwind of urban environments (see Chapter 9). TFA is a conversion product of the replacements for chlorofluorocarbons (CFCs), and an industrial solvent for Teflon™. An interesting aspect of rain is that raindrops will often carry an electric charge (i.e., a mix of positively and negatively charged raindrops (Gunn and Devin, 1953)). The electric charge may influence the nature of the absorption process.

### 3.4.2 Biota

When biota (e.g., fish, plants, and humans) are exposed to a chemical contaminant in water or air, the contaminant is taken up at a certain rate and eliminated at a certain rate. The bioconcentration index (BCI), defined as follows, can describe this process:

$$BCI = k_1 \text{intake} / k_2 \text{output} \qquad (3.34)$$

This allows the prediction of the BCF

$$BCF = C_o / C_{w,a} \qquad (3.35)$$

where $C_o$ is the concentration in the organism and $C_{w, a}$ is the concentration in water or air at equilibrium. Chemical contaminants can enter an organism through tissue penetration, ingestion, and by respiration. Penetration of tissue (e.g., human and animal skin; plant leaves and stems) by a chemical depends on its polarity (lipophilic or hydrophilic), molecular weight, charge, concentration, the area and duration of exposure, and species differences among plants and animals.

### 3.4.2.1 Biota and Water

Chemical contaminants in water are often in contact with plants, other organisms, and sediment. Use is made of the affinity of chemical contaminants in water to be absorbed by plants and adsorbed onto clays to create managed wetlands for water decontamination. Contaminated water is allowed to flow into the wetlands, where contaminants are removed and relatively clean water is released for more productive uses, such as fisheries or migratory bird sanctuaries. This principle has been suggested or put into practice in areas like the Chesapeake Bay, where surrounding intense farming or livestock production practices release undesirably high loads of nutrients to the Bay, encouraging algal blooms or contamination levels unsafe for fish and other aquatic species. Use of vegetative barriers or managed wetlands may be the most practical solution in such cases.

### 3.4.2.2 Root Zone Absorption

In general, water-soluble neutral organic molecules do not adsorb tightly to soil—the organic fraction or the clay mineral fraction. Examples include glyphosate, fluorinated detergents, TFA, phenoxy acetic acids, other water-soluble xenobiotics including heavy metals, and ionized or ionizable organics. This means that water-soluble insecticides will be available for absorption by plant roots and translocated, acting as systemic insecticides. Seiber et al. (1986) studied the absorption of root zone-applied systemic insecticides in rice paddies (Seiber et al., 1986) (see Chapter 10). They found that, after translocation, some of the insecticides were excreted in the guttation fluid and, once on the foliage surface, were available for volatilization. This is a mechanism whereby water-soluble systemics of low Henry's constant get into the air, rather than by direct water-to-air transfer, and become transportable airborne contaminants. Other soil constituents can be absorbed by plant roots and translocated. This may explain recent reports of arsenic, cadmium, and other heavy metals in rice, grapes, and other food products. Also, root zone absorption can be used to remediate/clean up contaminated soil. An example is the remediation of soil contaminated with radionuclides that resulted from past atomic testing (Entry et al., 1996).

### 3.4.2.3 Biota and Vapor

An approach analogous to the managed wetlands can be taken to clean the air of contaminants from around factories, feedlots, freeways, etc. Aston and Seiber (1997) followed the fate of pesticides in air from the large agricultural area in the San Joaquin Valley as the residues moved downwind and up gradient to the more pristine Sierra Nevada Mountains. They postulated that the heavily forested slope leading to the higher elevations would scrub pesticides from the air. This means that a volume of air downwind from the agricultural area would have less contaminant than what might be expected if only dilution were occurring. That is, some airborne chemical could be partitioned or absorbed by the pine needles. A cross section of a pine needle shows a waxy cuticle not unlike the coating on a solid-phase microextraction (SPME) device commonly used to sorb contaminants from air and water (see Chapter 6). The net result is that the air is "cleaned" following passage through evergreen forests. Forest buffers around factories and in urban parks (e.g., New York's Central Park) may play important roles in purifying the air (e.g., capturing excess carbon dioxide from combustion sources). People have increasingly come to recognize the value of planting trees or allowing some agricultural fields to be located around cities and transportation corridors.

## 3.5 Reactivity

In the summer months, when sunlight is its most intense and skies are relatively cloudless, pesticide application is heaviest. This is significant because light is the main factor driving pesticide transformations in air (Woodrow et al., 2018). Theoretically, this makes sense (Woodrow et al., 1983): The bond energies in volatile organic chemicals are in the range of energies of wavelengths of sunlight. Direct reactions with air occur at wavelengths from 290 to 400 nm. Considering the energy at each wavelength:

$$E = Lhc/\lambda \qquad (3.36)$$

where L is Avogadro's number ($6.022 \times 10^{23}$ molecules/mol), h is Planck's constant ($6.626 \times 10^{-27}$ erg/sec), and c is the speed of light ($3 \times 10^{10}$ cm/sec). E, energy (kcal/mol), of the C–C bond in ethane is 88 kcal/mol and for a C–H bond in the same molecule is 98 kcal/mol, while the energies of sunlight in the lower atmosphere are from 96 to 72 kcal/mol at wavelengths ($\lambda$) from 300 to 400 nm. In the stratosphere, lower wavelengths are present that can catalyze activation of even stronger bonds (see Chapter 9). Many pesticides absorb sunlight from 290 to 400 nm and readily undergo photocatalyzed reactions in air. Indirect reactions also occur in air in which singlet and triplet oxygen, hydroxyl radicals, ozone, and nitrogen oxides transfer the energies needed to break bonds in atmospheric pesticides and toxics (Atkinson et al., 1999; Woodrow et al., 2018). The frequencies of these indirect reactions are dependent on air quality. Particulate load and moisture in the air also influence reactions (Epstein and Nizkorodov, 2012; Socorro et al., 2017). The reactions in air affect the atmospheric lifetime of a pesticide and thus its long-range transfer to remote regions, its impacts on the atmosphere, and its toxicity.

Parathion applied to orchards in the midday in summertime is oxidized to its more toxic counterpart paraoxon in minutes (Woodrow et al., 1977). Higher ratios of paraoxon to parathion downwind of the application site confirmed the chemical conversion of parathion to paraoxon in air is a major transformation route. Eventually, parathion is converted to the breakdown product, p-nitrophenol. Conversion of parathion to paraoxon is accelerated by sunlight, but slower conversion does occur at night (half-life during day is 2 minutes, but at night it is 131 minutes) (Woodrow et al., 1978). Similar experiments were used to measure transformation of chlorpyrifos to its more stable reactant product chlorpyrifos oxon (Aston and Seiber, 1997). As the elevation increased, so did the levels of chlorpyrifos oxon compared to the parent, indicating that atmospheric oxidation is a dominant pathway.

Chambers in the laboratory can also be used to study photoinitiated reactions using lamps that simulate sunlight. Trifluralin photodecomposed to multiple products in the lab in a vapor phase reactor equipped with an air sampler collecting vapors on a coated solid sorbent with limit of detection of less than 1 $ng/m^3$ (Soderquist et al., 1975). The half-life of trifluralin was 47 minutes in the lab and between 21 and 193 min in the field, showing good agreement. Photolysis occurring at the soil surface followed by volatilization was the dominant occurrences. Trifluralin is stable in the dark (Woodrow et al., 1978).

Sulfur is applied at the largest volume rate as a pesticide worldwide including in California (see Table 2.2) (Griffith et al., 2015). The dissipation routes of sulfur from foliage include wind erosion, washing off, microbial activity, oxidation, and photooxidation. At 39°C, sulfur has a vapor pressure of $1.33 \times 10^{-3}$ Pa, corresponding to a loss of 449 $\mu g/m^2 \cdot hr$ assuming little or no interaction with plant surfaces soon after application (Woodrow et al., 2001). There are no direct studies of the atmospheric reactions of agricultural sulfur; however, it is expected it would undergo the same oxidation transformation reactions as other sulfur species.

Volatile organic compounds partition to vapor phase (fog and clouds) where they undergo aqueous phase photolysis reactions and reactions with hydroxyl radicals (Kaur and Anastasio (2017). Epstein and Nizkorodov (2012) used complex modeling to evaluate photolysis reactions for several classes of organics and point out that this is another significant pathway for pesticides in air. The importance of aqueous-phase photolysis can be assessed by the propensity of the chemical to partition to the fog or cloud droplets from air (see Chapter 7).

Some of the major reactions of pesticides include N- and O-dealkylations, exemplified by trifluralin N-dealkylation; photolytic decomposition, exemplified by methyl isothiocyanate to the more toxic and more stable methyl isocyanate (see Chapter 8); and oxidations: mercaptans and disulfides to $SO_2$, thions to oxons, and merphos (S,S,S-tributyl phosphorotrithioite) to its trithioate conversion product Def (Woodrow et al., 2018). Polycyclic aromatic hydrocarbons (PAHs) such as dibenzanthracene that are released during plant biomass burning are activated to epoxide forms that are prominent mutagens and carcinogens (Chapter 4). But photoreactions of pesticides in fog and air represents an area which warrants further work in terms of mechanisms and rates.

Reactions of pesticides in air, as well as in moisture and particulate matter in air, are not negligible and require considerations that must be studied by companies during pesticide registration. Understanding these reactions has led to discontinued use of some heavily used pesticides, and it helps understand the fate of pesticides and their long-range transport. The implications of this are discussed more in Chapter 10.

## 3.6 Conclusions

No environmental compartment (water, soil, air, biota) exists in isolation. After an intentional or accidental release, a chemical contaminant will be distributed to varying degrees among the various compartments, depending on the properties of the chemical (organic content, water solubility, vapor pressure, diffusion coefficient, BCF) and the environmental conditions. In these different compartments, transformations may occur for reactive chemicals. The concepts covered in this chapter should contribute to a better understanding of the environmental fate of chemical contaminants, leading to a more confident answer to the question "where did that chemical go?" Clearly more work is needed, particularly on transformation of pesticides and other contaminants, including viruses in air.

## References

Angermann, J.E., Fellers, G.M., and Matsumura, F. (2002). Polychlorinated biphenyls and toxaphene in Pacific tree frog tadpoles (Hyla regilla) from the California Sierra Nevada, USA. *Environ. Toxicol. Chem.* 21(10), 2209–2215. DOI: 10.1002/etc.5620211027.

Aston, L.S., and Seiber, J.N. (1997). Fate of summertime airborne organophosphate pesticide residues in the Sierra Nevada Mountains. *J. Environ. Qual.* 26(6), 1483–1492. Doi: 10.2134/jeq1997.00472425002600060006x.

Atkinson, R., Guicherit, R., Hites, R.A., Palm, W., Seiber, J.N., and de Voogt, P. (1999). Transformations of pesticides in the atmosphere: A state of the art. *Water, Air, Soil Pollut.* 115(1), 219–243. Doi: 10.1023/A:1005286313693.

Entry, J.A., Vance, N.C., Hamilton, M.A., Zabowski, D., Watrud, L.S., and Adriano, D.C. (1996). Phytoremediation of soil contaminated with low concentrations of radionuclides. *Water. Air. Soil Pollut.* 88, 167–176.

Epstein, S.A., and Nizkorodov, S.A. (2012). A comparison of the chemical sinks of atmospheric organics in the gas and aqueous phase. *Atmosp. Chem. Phys.* 12, 8205–8222.

Gawlik, B., Sotiriou, N., Feicht, E., Schulte-Hostede, S., and Kettrup, A. (1997). Alternatives for the determination of the soil adsorption coefficient, KOC, of non-ionic organic compounds—a review. *Chemosphere* 34, 2525–2551.

Glotfelty, D.E., Seiber, J.N., and Liljedahl, A. (1987). Pesticides in fog. *Nature* 325, 602–605. Doi: 10.1038/325602a0.

Griffith, C.M., Woodrow, J.E., and Seiber, J.N. (2015). Environmental behavior and analysis of agricultural sulfur. *Pest Manage. Sci.* 71(11), 1486–1496. Doi: 10.1002/ps.4067.

Gunn, R., and Devin Jr, C. (1953). Raindrop charge and electric field in active thunderstorms. *J. Meteorol.* 10, 279–284.

Kaur, R., and Anastasio, C. (2017). Light absorption and the photoformation of hydroxyl radical and singlet oxygen in fog waters. *Atmos. Environ.* 164:387–397.

Lewis, W.K., and Whitman, W.G. (1924). Principles of gas absorption. *Indust. Eng. Chem.* 16(12), 1215–1220. Doi: 10.1021/ie50180a002.

Lyman, W.J. (1985). Estimation of physical properties. *Environ. Expo. Chem.* 1, 30–31.

Mackay, D., and Boethling, R.S. (2000). *Handbook of Property Estimation Methods for Chemicals: Environmental Health Sciences.* Boca Raton: CRC Press.

Mackay, D., and Van Wesenbeeck, I. (2014). Correlation of chemical evaporation rate with vapor pressure. *Environ. Sci. Technol.* 48(17), 10259–10263. Doi: 10.1021/es5029074.

Majewski, M.S., Glotfelty, D.E., and Paw U.K.T. (1990). A field comparison of several methods for measuring pesticide evaporation rates from soil. *Environ. Sci. Technol.* 24(10), 1490–1497. Doi: 10.1021/es00080a006.

Majewski, M.S., Glotfelty, D.E., and Seiber, J.N. (1989). A comparison of the aerodynamic and the theoretical-profile-shape methods for measuring pesticide evaporation from soil. *Atmos. Environ.* 23(5), 929–938. Doi: 10.1016/0004-6981(89)90297-7.

Majewski, M.S., McChesney, M.M., Woodrow, J.E., Prueger, J.H., and Seiber, J.N. (1995). Aerodynamic measurements of methyl bromide volatilization from tarped and nontarped fields. *J. Environ. Qual.* 24(4), 742–752. Doi: 10.2134/jeq1995.00472425 002400040027x.

Ott, R. (2005). *Sound Truth and Corporate Myth$: The Legacy of the Exxon Valdez Oil Spill.* Cordova, Alaska: Dragonfly Sisters Press.

Rolston, D.E. (1986). Gas Flux. In A. Klute (Ed.), *Methods of Soil Analysis Part 1: Physical and Mineralogical Methods* (2nd ed., pp. 1103–1119). Doi: 10.2136/sssabookser5.1. 2ed.c47.

Seiber, J.N., Madden, S.C., McChesney, M.M., and Winterlin, W.L. (1979). Toxaphene dissipation from treated cotton field environments: component residual behavior on leaves and in air, soil, and sediments determined by capillary gas chromatography. *J. Agric. Food Chem.* 27, 284–291.

Seiber, J.N., McChesney, M.M., Sanders, P.F., and Woodrow, J.E. (1986). Models for assessing the volatilization of herbicides applied to flooded rice fields. *Chemosphere* 15(2), 127–138. Doi: 10.1016/0045-6535(86)90564-3.

Seiber, J.N., and Woodrow, J.E. (1983). Methods for studying pesticide atmospheric dispersal and fate at treated areas. In F.A. Gunther, and J.D. Gunther (Eds.), *Residue Reviews* (Vol. 85, pp. 217–229). New York. Doi: 10.1007/978-1-4612-5462-1_16.

Socorro, J., Lakey, P.S., Han, L., Berkemeir, T., Lammel, G., Zetzch, C. et al., (2017). Heterogeneous OH oxidation, shielding effects, and implications for the atmospheric fate of terbuthylazine and other pesticides. *Environ. Sci. Technol.* 51(23), 13749–13754. Doi: 10.1021/acs.est.7b04307.

Soderquist, C.J., Crosby, D.G., Moilanen, K.W., Seiber, J.N., and Woodrow, J.E. (1975). Occurrence of trifluralin and its photoproducts in air. *J. Agricult. Food Chem.* 23(2), 304–309. Doi: 10.1021/jf60198a003.

Thornthwaite, C.W., and Holzman, B. (1939). The role of evaporation in the hydrologic cycle. *Eos, Transact. American Geophys. Union* 20(4), 680–686. Doi: 10.1029/TR020i004p00680.

Wilson, J.D., Thurtell, G.W., Kidd, G.E., and Beauchamp, E.G. (1982). Estimation of the rate of gaseous mass transfer from a surface source plot to the atmosphere. *Atmosp. Environ.* 16(8), 1861–1867. Doi: 10.1016/0004-6981(82)90374-2.

Woodrow, J.E. (2003). The laboratory characterization of jet fuel vapor and liquid. *Energy Fuels* 17(1), 216–224. Doi: 10.1021/ef020140p.

Woodrow, J.E., Crosby, D.G., Mast, T., Moilanen, K.W., and Seiber, J.N. (1978). Rates of transformation of trifluralin and parthion vapors in air. *J. Agri. Food Chem.* 26(6), 1312–1316. Doi: 10.1021/jf60220a019.

Woodrow, J.E., Crosby, D.G., and Seiber, J.N. (1983). Vapor-phase photochemistry of pesticides. In F.A. Gunther, and J.D. Gunther (Eds.), *Residue Reviews* (Vol. 85). New York: Springer. Doi: 10.1007/978-1-4612-5462-1_9.

Woodrow, J.E., Gibson, K.A., and Seiber, J.N. (2018). Pesticides and related toxicants in the atmosphere. In P. de Voogt (Ed.), *Reviews of Environmental Contamination and Toxicology* (Vol. 247, pp. 147–196). Cham: Springer International Publishing.

Woodrow, J.E., LePage, J.T., Miller, G.C., and Hebert, V.R. (2014). Determination of methyl isocyanate in outdoor residential air near metam-sodium soil fumigations. *J. Agri. Food Chem.* 62(36), 8921–8927. Doi: 10.1021/jf501696a.

Woodrow, J.E., and Seiber, J.N. (1991). Two chamber methods for the determination of pesticide flux from contaminated soil and water. *Chemosphere* 23(3), 291–304. Doi: 10.1016/0045-6535(91)90185-G.

Woodrow, J.E., Seiber, J.N., and Baker, L.W. (1997). Correlation techniques for estimating pesticide volatilization flux and downwind concentrations. *Environ. Sci. Technol.* 21(2), 523–529. Doi: 10.1021/es960357w.

Woodrow, J., Seiber, J.N., Crosby, D., Moilanen, K., Soderquist, C., and Mourer, C. (1977). Airborne and surface residues of parathion and its conversion products in a treated plum orchard environment. *Arch. Environ. Contam. Toxicol.* 6, 175–191.

Woodrow, J.E., Seiber, J.N., and Dary, C. (2001). Predicting pesticide emissions and downwind concentrations using correlations with estimated vapor pressures. *J. Agric. Food Chem.* 49, 3841–3846.

Woodrow, J.E., Seiber, J.N., and Kim, Y.-H. (1986). Measured and calculated evaporation losses of two petroleum hydrocarbon herbicide mixtures under laboratory and field conditions. *Environ. Sci. Technol.* 20(8), 783–789. Doi: 10.1021/es00150a004.

Woodrow, J. E., Seiber, J.N., and Miller, G.C. (2011). Correlation to estimate emission rates for soil-applied fumigants. *J. Agri. Food Chem.* 59(3), 939–943. Doi: 10.1021/jf103868k.

Yates, S.R. (2006). Measuring herbicide volatilization from bare soil. *Environ. Sci. Technol.* 40(10), 3223–3228. Doi: 10.1021/es060186n.

# 4

# *Pesticide Exposure and Impact on Humans and Ecosystems*

## 4.1 Introduction

Most of what we know regarding exposure to pesticides is from water and food intake. Inhalation and other forms of air exposure have been less studied. This may be due to the fact that airborne exposures are infrequent or of limited durations that vary with place and time of day. However, airborne exposure is a principal means of exposure for particulate pesticides including dust and those applied as fine dust, such as sulfur, and exposure by the air is a principal means of transmission of viruses like SARS-CoV-2, the virus that causes the coronavirus disease of 2019 (COVID-19). The principal means of safeguard is through use of masks and social distancing. Thus, the airborne route of exposure is gaining new attention.

## 4.2 Examples of Human Exposures

### 4.2.1 Paraquat

There is current interest in a potential link between exposure to paraquat, a leading herbicide in agricultural and home use weed control, with Parkinson's disease (Butcher, 2020). Due to its neurotoxicity, many countries have banned paraquat; however, it is still widely used in the United States, with estimates that the usage doubled from 2006 to 2016 (Butcher, 2020). Exposure to paraquat and to trichloroethylene, a leading dry-cleaning agent and drinking water contaminant, are suspects in the increase in Parkinson's disease, which has increased in case load by four times in the United States in recent years—leading some to conclude that Parkinson's disease is a "man-made" disease.

DOI: 10.1201/9781003217602-4

Exposure to paraquat could occur by inhalation of aerosolized spray during application (Woodrow et al., 1989). Anecdotal evidence has been offered in support of this hypothesis—yet to be proven with controlled studies. Paraquat was once widely used, in a joint effort by Mexico and the United States in the 1990s, to destroy marijuana plants to reduce availability of then illegal "weed" or "pot." Yet another source of exposure was believed to be from inhaling smoke from cannabis cigarettes, assuming that cannabis could retain paraquat in sprayed marijuana that did not succumb to the spray. An agricultural link might have been exposure to spray by applicators of paraquat during its use, widespread in California and some southern states and in Arizona, to defoliate cotton prior to harvest of cotton lint and for weed control. Defoliation was used to reduce the load of cotton gin trash and improve quality of the lint harvested during spindle picking of cotton lint.

### 4.2.2 Petroleum-Based Weed and Dormant Oil Pesticides

Petroleum oils are commonly used to control weeds using primarily diesel and mineral oil fractions (see Table 2.2). This practice is now discouraged because these petroleum fractions can lead to formation of photochemical smog (San Joaquin Valley Air Pollution Control District, 2012). Oils are also applied to new crops and as harvest aids (Woodrow et al., 1986). These oils can volatilize with lower-molecular-weight fractions volatilizing first giving way to residue of heavier hydrocarbons which then dissipate by other means (e.g., microbial metabolism).

Spills of crude oil occur during accidents, and the risks of spills increase during hurricanes and other natural disasters. For example, the Deepwater Horizon spill, caused by multiple failures while closing up the well, spilled more than 700,000 tons to the Gulf of Mexico. Prior to that, 27,000–37,000 tons of oil were released during hurricane Katrina from various sources in the Louisiana parish that is heavily involved in the oil industry.

In both events, it is estimated that half of the spilled oil evaporated (a familiar thought about things disappearing in the vast air) (Farrington, 2013). In these cases, oil was spilled in the relatively warm water in the Gulf of Mexico (NCEI NOAA, 2018b; U.S. EPA, 2015a). The warm water (up to 32°C in summer months) plus the windy conditions led to high evaporation rates. The 1989 Exxon Valdez oil spill occurred in Prince William Sound, Alaska, a region of cooler water with an average high of 14°C in summer months (NCEI NOAA, 2018a). The spilled oil did not evaporate as much initially as the spills in the warmer Gulf of Mexico, but overtime loss to the air by volatilization and other dissipation mechanisms was significant. Also, the Deepwater Horizon accident released a lighter volatile crude oil compared to the heavy crude oil released in the Exxon Valdez incident in Alaska (Atlas and Hazen, 2011). The lighter crudes are more volatile and more readily biodegradable. Plus, in the Gulf of Mexico, microorganisms are adapted to degrading hydrocarbons (Atlas and Hazen, 2011).

### 4.2.3 Other Examples

One example of an acute exposure episode involving a toxic volatile industrial chemical was the accidental release of methyl isocyanate gas (an intermediate in the production of carbamate pesticides) from a chemical plant in Bhopal, India, in 1984 (Varma and Varma, 2005). The released gas engulfed a nearby village resulting in thousands of fatalities and injuries.

Another example involved workers involved in the construction of a tunnel linking Sweden with Denmark. The tunnel walls were grouted with acrylamide, which polymerized to polyacrylamide to seal the tunnel against intrusion of moisture. During grouting, the unpolymerized acrylamide monomer was released inadvertently to the air at levels resulting in exposures to hundreds of workers at significant levels of the neurotoxin (Kjuus, 2001). Ironically, the same chemical (acrylamide) is formed via the Maillard browning reactions that occur during cooking of high carbohydrate foods, resulting in population-wide low-level exposures to acrylamide, a carcinogen and cumulative neurotoxin (Tareke et al., 2002).

### 4.2.4 Mitigation

Public awareness of the potential risks of pesticide exposure has prompted pesticide manufacturers and applicators to decrease exposures to nearby populations. It is particularly important for local authorities and governments to enact measures that would protect nearby residents and ecosystems. Pesticide manufacturers and applicators have responded by (1) using salts of pesticides that can be applied via drip irrigation or other targeted means (as opposed to the spray planes) (Seiber et al., 2014); (2) use of biopesticides that have a more targeted effect and generally break down in the environment faster (Chapters 2 and 12); (3) the use of pheromones or semiochemicals, for example, to trap insects; and (4) the use of genetically modified crops that make their own pesticides. In combination, these strategies can significantly reduce exposures to communities in proximity to high agricultural activities, especially in California, the Midwest United States, and Florida.

## 4.3 Regulations and Classifications

Regulatory agencies that limit pesticide exposure include international agencies (World Health Organization (WHO)), regional and national agencies (European Union (EU), Environmental Protection Agency (EPA), Food and Drug Administration (FDA)), statewide agencies (California Department of Pesticide Regulation (CDPR)), and individual counties and municipalities. These agencies are described in the accompanying glossary. For the most part, they focus on the residue in food and water and not on residues in air. Also, there are no enforceable regulatory limits to amounts permissible for dermal exposure, just general statements advising to avoid and wash off when formulations are spilled on clothing or body.

Some regulations on contaminants in air are discussed in this section. There are no enforceable national or international limits describing a permissible level of exposure to toxics in air applicable to ambient populations, however, the U.S. Occupational Safety and Health Administration (OSHA) issues guidance on threshold limit values (TLVs) for many chemicals including some pesticides.

Beside pesticides, there are the other airborne chemicals of concern. Termed criteria air pollutants and regulated by the Clean Air Act (U.S. EPA), they include particulate matter, photochemical oxidants (e.g., ozone ($O_3$)), carbon monoxide (CO), sulfur oxides (SOx), nitrogen oxides (NOx), hydrocarbons (HxCy), and lead. EPA sets National Ambient Air Quality Standards (NAAQs) based on the latest scientific information (Griffith et al., 2015; Laumbach and Kipen, 2012; Seiber et al., 1979, 2014; U.S. EPA, 2021).

Chemical manufacturing plants are a source of toxic chemicals to the air. Examples include chlorinated solvents, monomers, and the organic contaminants in gasoline known as BETX (benzene and ethyl benzene, xylene, toluene). These compounds are also classified as priority pollutants by EPA because they can contaminate groundwater, surface water, and drinking water. Because they have high Henry's law constants, they often wind up in the air. They are of concern due to their acute and chronic effects on exposed populations (e.g., workers, emergency responders, nearby residents). The chemicals listed in Table 4.1 can contaminate drinking water and readily undergo exchange between air, water, and surfaces due to their high Henry's law constants and relatively low water solubility.

Guidelines for exposure level of volatile organic compounds (VOCs) are given by the American Council of Governmental Industrial Hygienists (ACGIH) and various other regulatory agencies. VOCs are sometimes termed "Purgeable Organics" (Lopez-Avila et al., 1987). U.S. EPA also classifies some pesticides as hazardous air pollutants (HAPs). In California, many pesticides are also classified as toxic air contaminants (TAC) under AB 1807. Between AB 1807 and California's unique proposition 65, pesticides and other chemicals received multiple regulatory limitations (CARB, 2017). Of the TACs in California, 16 are pesticides. At the international level, persistent organic pollutants (POPs) are known to undergo transport in the air to more distant areas via long-range transport (see Table 4.2).

## 4.4 Indoor vs Outdoor Chemicals

People spend most of their time indoors, in houses, buildings, stores, etc. Thus, attention to indoor chemical exposure is warranted. The trend toward more

**TABLE 4.1**

Properties of Representative VOCs and SVOCs

| VOC | Vapor Pressure at 25°C (Pa) | Henry's Law Constant (Pa m³/mol) | Water Solubility (mg/L) | IARC Classification[a] | TAC | HAP[b] | Prop 65 |
|---|---|---|---|---|---|---|---|
| *Solvents, Aromatic Hydrocarbons (HCs), Nonchlorinated* | | | | | | | |
| Benzene | 1.26E04 | 562.3 | 1,880 | 1 | | x | x |
| Ethylbenzene | 1.28E03 | 798.4 | 157.63 | 2B | | | x |
| Toluene | 3.79E03 | 672.8 | 526 | 3 | | x | x |
| Xylene | 8.81E02 | 681.9 | 161.92 | 3 | x | | x |
| *Polycyclic Aromatic HCs* | | | | | | | |
| Naphthalene | 1.13E01 | 48.9 | 31.0 | 2B | x | x | x |
| Dibenz[a,h]anthracene | 1.27E-07 | 1.5E-03 | 0.00249 | 2A | x | x | |
| *Pesticides* | | | | | | | |
| Chlorpyrifos | 2.71E-03 | 1.114 | 1.4 | N/A | x | | x |
| Endosulfan | 2.31E-05 | 1.135 | 0.510 | N/A | x | | |
| Lindane | 4.70E-03 | 1.418 | 6.80 | 1 | x | x | x |
| *Pesticides (Fumigants)* | | | | | | | |
| Methyl bromide | 2.16E05 | 632.3 | 1.52E04 | 3 | x | x | x |
| Methyl isothiocyanate (MITC) | 2666 | 0.679 | 7,600 | N/A | x | x (MIC) | x (MIC) |
| 1,3-Dichloropropene (Telone) | 4.53E03 | 1,793.4 | 2,800 | 2B | x | x | x |
| *Monomers* | | | | | | | |
| Vinyl chloride | 3.97E05 | 2,735.7 | 2,760 | 1 | | x | x |
| Acrylonitrile | 1.45E04 | 10.13 | 74,000 | 2B | | x | x |
| Styrene | 8.53E02 | 2.786 | 310 | 2B | | x | x |

*Source:* Adapted from Woodrow et al. (2018).

[a] IARC classification: Group 1, carcinogenic to humans; Group 2A, probably carcinogenic to humans; Group 2B, possibly not carcinogenic to humans; Group 3, not classifiable as to its carcinogenicity to humans; Group 4, probably not carcinogenic to humans (IARC 2018).

[b] U.S. EPA (2016). HAP list.

**TABLE 4.2**

POPs aka the "Dirty Dozen"

| POP | Uses and Occurrences | Usage Cancelled | HAP |
|---|---|---|---|
| Aldrin and dieldrin | Insecticide (agricultural, urban) | 1987 | |
| Chlordane | Insecticide (agricultural, urban) | 1988 | x |
| Heptachlor | Insecticide (agricultural, urban) | 2000 | |
| Hexachlorobenzene | Fungicide, industrial, unintentional (combustion, by-product, trace impurity) | 1985 (pesticide use), industrial uses allowed | x |
| PCBs | Industrial, unintentional in combustion | 1978 | x |
| Dioxins and furans | Unintentional (combustion, by-product, trace impurity) | | x |

*Source:*   Kovner (2002).

energy efficient buildings over recent decades has created buildings with reduced ventilation from outdoors; attention now should also include design of healthy buildings (Ghaffarianhoseini, 2018). Indoor pollutants come from oil, gas, tobacco products, carpets, common household cleaning products, and heating, ventilation, and air conditioning (HVAC) systems (U. S. EPA, 2015b). Tightly sealed homes have lower ventilation rates from outdoors that can lead to mucosal symptoms, especially at night in the winter (Sun et al., 2019).

There is a growing realization that people and their pets track in dirt and other contaminants including pesticides to inside houses (Williams et al., 2004). This is additive to the pesticides intentionally used in homes to kill rodents, roaches, termites, and other pests. Several of the "dirty dozen" chemicals are pesticides with urban and household uses that have been banned, including chlordane, dichlorodiphenyltrichloroethane (DDT), and heptachlor. Passive air sampling in homes in China found polycyclic aromatic hydrocarbons (PAHs) and the pesticide hexachlorobenzene (a "dirty dozen" chemical) from outdoors, and chlordane and pyrethrins from pest control indoors (Wang et al., 2019). Exposure to chlordane has been linked to testicular dysgenesis (Cook et al., 2011), diabetes (Evangelou et al., 2016), obesity (Tang-Péronard et al., 2011), and cancers (Khanjani et al., 2007; Lim et al., 2015; Luo et al., 2016). Polychlorinated biphenyls (PCBs), also included in the "dirty dozen," were another indoor source of contaminants that were especially high in older homes (Wang et al., 2019). Collectively, these indoor exposures lead to "sick building syndrome" (Wargocki, 2000). "Sick building syndrome" is also experienced in schools due to poor ventilation and high temperatures; student performance significantly improved when schools improved the indoor air quality (Wargocki, 2006). Current attention is focused on COVID-19 in indoor air (see Chapter 11).

Again, no universal standards exist for pesticides and other toxic contaminants in ambient air.

## 4.5 Human Exposure Routes and Impacts

The simple act of breathing exposes human populations to pesticides and other airborne contaminants (Woodrow et al., 2018). For example, field workers spraying 2.5% aqueous chlorothalonil are exposed to 4 μg of the chemical per hour (Dowling and Seiber, 2002). Inhalation toxicity of many chemicals is not known, especially at low concentration exposures experienced chronically (Woodrow et al., 2018). Studies are particularly important for sensitive populations: children and fetuses (since many toxics have the strongest impacts during development), the elderly (who often spend most of their time indoors), and economically disadvantaged people (more likely to live near industrial zones and waste management systems). Areas with historically high particulate matter exposures have higher mortality rates due to COVID-19 (Wu et al., 2020). To properly study the impacts of breathing pesticides and toxics, the rates of inhalation of different populations engaged in different activities must be known. The EPA has a review of the inhalation rates of people by age and level of exertion (U.S. EPA, 2011).

Inhaled vapors and particle matter with adsorbed chemicals, are absorbed by the lungs and enter the bloodstream, thus reaching all organs and tissues (Ngo et al., 2010). The size of particulate matter (PM) affects its release after inhalation, as well as depth of lung penetration. PM10 is effectively removed in the nasal passageway (Table 4.3). PM2.5 and lower penetrate the lungs: PM2.5 are deposited in tracheobronchial and alveolar regions while PM0.1 reach the alveolar (low lung) region. Thus, there are different health-driven regulatory approaches toward PM of varying sizes.

PM from combustion of plant material is emitted to air from sources that are preventable (smoking, agricultural residue burning, combustion of wood in fireplaces) and from accidental/natural disasters such as grassland and

**TABLE 4.3**

Particle Size of Common Indoor Air Contaminants (Air Quality Sciences, 2011; CCOHS, 1997)

|  | Size Range and Penetration | Examples with Sizes (μm) |
|---|---|---|
| PM10 | 10 μm or larger; are trapped and removed efficiently in the nasal region | Visible lint (>25); resuspended dust (5–25); some mold and pollen spores (2–200) |
| PM2.5 | 0.003–5 μm; are deposited in the tracheobronchial and alveolar regions | Some mold and pollen spores (2–200); environmental tobacco smoke (0.1–0.8); diesel soot (0.01–1); outdoor fine particles (sulfates, metals) (0.1–2.5) |
| PM0.1 | 0.5 μm or smaller; are deposited in the alveolar regions (lower lung area) | Viruses (<0.01–0.05); bacteria (0.05–0.7); fresh combustion particles (<0.1); ozone- and terpene-formed aerosols (<0.1) |

*Source:* Adapted from Woodrow et al. (2018).

forest fires. The devastating wildfires of 2020 in the Western United States are thrusting the hazards of PM from smoke to the fore; smoke from wildfires will likely carry a range of toxins, similar to those in the by-products of agricultural burning. Cigarette smoke and marijuana smoke (legalized in many places in the United States) are a source of burned plant materials in and around the home. Because it is illegal in most places, and only newly legal in some places, marijuana receives less pesticide monitoring than tobacco (Taylor and Birkett, 2020). Common pesticides used to grow marijuana include insecticides, acaricides, and fungicides. Though amounts were not reported, ~85% of tested cannabis samples had pesticides in Washington State, and ~50% in California.

Agricultural burning was a common practice in California but has been banned (Figure 4.1). However, it is still common practice in other places including China and has reportedly increased over the last decade (Wang et al., 2020). Rice straw releases PAH, phenols, and mutagens similar to cigarette smoke (Mast et al., 1984). PAHs released by burning plant material can be oxidized to form even more dangerous "activated" products. For example, dibenzanthracene is activated by oxidation to form epoxides that are prominent mutagens and carcinogens. More research is needed

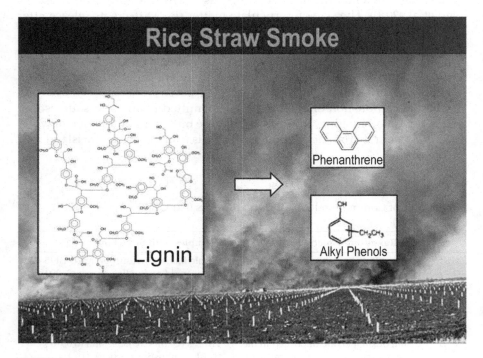

**FIGURE 4.1**
Agricultural burning, for example, of rice straw is a common agricultural problem throughout much of the world.

to advise regulatory agencies in places where agricultural burning is a common practice. The inhalation of aerosols bearing toxins, viruses, and pathogens follows the same principles. COVID-19 has triggered a new level of research into exposures to pathogenic microorganisms via aerosols.

## 4.6 Exposures to Ecosystems

Once contaminants enter the air, they are out of our control (Woodrow et al., 2018). They can travel in the wind and be dispersed to other places that may be great distances from the source, often in fragile ecosystems (see Chapter 10). Even when detected at trace levels, deposited chemicals impact the quality of waters, in water bodies such as lakes, rivers, oceans including snow and ice and water sheds. Oceanic residues of organochlorine pesticides that undergo bioaccumulation affect top predators, for example, DDT at a concentration below one part per billion in water can result in higher contamination in fish (2 ppm) and fish-eating birds (25 ppm) (Woodrow et al., 2018). The same goes for other nonpolar contaminants such as PCBs and some mercury- and selenium-containing contaminants.

In the 1970s, there were frequent reports of sick or dying red-tailed hawks largely in the orchards in the central and southern San Joaquin Valley of California. A collaborative study was undertaken involving the research team of Professor Barry Wilson of the University of California, Davis, Department of Avian Science, Jim Seiber's team in the Department of Environmental Toxicology, and others. Wilson approached Seiber with a proposition: Can we do analysis of hawks to see if they are being exposed to toxic levels of organophosphate (OP) pesticide? These teams analyzed hawk feathers, talons, and excreta for OPs. They found residues of the OPs parathion and a few others. But surprisingly high ("off the chart") levels of paraoxon, a highly toxic environmental conversion product of parathion, were found in hawk feathers and in orchard air. Was paraoxon the possible culprit? Wilson thought so, but it was not until his group "connected the dots" and calculated exposure to paraoxon the hawks would experience from feather preening. Samples from feathers were markedly higher in oxon than other samples. Thus, the toxic effects of oxons seen in orchards could be explained by hawk ingestion of oxons during preening. Carol Weiskopf, a grad student in Seiber's lab, used forensic techniques to find the diagnostic metabolites of paraoxon in stomachs and excreta of red-tail hawks (Wilson et al., 1991).

Colony collapse disorder among honeybees is of current concern. It results when adult worker bees die in and around the hive (Oldroyd, 2007; vanEngelsdorp et al., 2009). Some suspect pesticides, especially neonicotinoids (see Chapter 2) (Gill et al., 2012; Kiljanek et al., 2016).

## 4.7 Global Distribution and Climate Change

Atmospheric transport is a major route for transporting pesticides globally (Woodrow et al., 2018). In the tourist island of Pingan (Fujian Province, China), seasonal variations in organochlorine (OC) pesticides correlated with total particulate levels, which tend to be higher in winter and lower in summer (Jiao et al., 2018). The source of the polluted air was traced to the air mass from the "heating season" in Northern China, known to be laden with particles. The same trend was observed in Jinan, China, from July 2009 to June 2010 for OC pesticides in particulate matter (Xu et al., 2011). High concentrations of OC pesticides were detected in winter through spring and were lower in summer. The long-range transport could be attributed to seasonal usage as well as meteorological conditions.

In the Great Lakes region of the United States, pesticide levels in air were greater in urban areas than in rural or remote sites, indicating pesticides in the air can be attributed to agricultural use as well as other human activities (Wang et al., 2018). Another study by the same group in the Great Lakes region looking at both seasonal and spatial variation of gas particle partitioning again found that current-use pesticides were higher in urban areas. Pyrethrins accounted for the high levels of atmospheric pesticides in urban areas, while fungicides were highest in agricultural regions. Relative humidity and temperatures did not influence vapor phase pesticides (e.g., trifluralin and chlorpyrifos), but the particle/aerosol phase of metolachlor was marginally lower at higher humidity. While median levels of the current use pesticides measured by Wang et al. (2018) were lower in remote regions, their presence confirms transport to remote regions in which pesticides are not intentionally applied.

Stable pesticides and stable degradates give important clues to the long-range transport of POPs (e.g., DDT, toxaphene, chlordane, and lindane). An expedition to the Arctic revealed higher relative levels of p, p'-DDT and o, p'-DDT in east Asia and the North Pacific indicating drift of recent use DDT and related compounds on the adjacent content (Wu et al., 2011). Climate change affected the distribution of enantiomers of $\alpha$-hexachlorocyclohexane (HCH) and the geometric isomers of chlordane in the Arctic (Bidleman et al., 2015). Based on the ratios of $\gamma$-HCH to $\alpha$-HCH, long-range transport is the main contributing factor for atmospheric POPs in Antarctica (and not nearby research stations) (Hao et al., 2019).

Global warming not only exacerbates the transport of pesticides and POPs to remoter regions—through atmospheric movement and ocean currents, but it also promotes secondary emissions of POPs from glaciers and permafrost (Nadal et al., 2015; Wang et al., 2016; Bidleman et al., 2015). In addition to POPs, other toxics like mercury are likely to be released from Arctic permafrost due to global warming (Schuster et al., 2018). To minimize the impact of these climate changes, more research is needed.

## 4.8 Conclusions

A future challenge is to assess and manage the effects of exposures to known or suspected toxicants in a comprehensive manner. Earlier generations of researchers focused on cancer, birth defects, AIDS, and other health outcomes. Age-related diseases like Alzheimer's disease have risen in interest, as has coronary heart disease and diabetes, which are associated with lifestyles including diet, overuse of alcohol, and poor diet–exercise choices.

Parkinson's disease is the fastest-growing neurological disease in terms of cases and deaths, and research suggests various environmental factors could be at work, including exposure from pesticides such as rotenone and paraquat (Chen and Ritz, 2018). As a synthetic chemist and one of millions of people afflicted with Parkinson's, author JNS, wonders if his exposures to solvents, synthetic product mixtures of often unknown composition, pesticides—particularly OCs such as toxaphene and OPs and carbamates, paraquat and phenoxy herbicides—have led to his current condition. Or perhaps there are other factors which have yet to be studied. Parkinson's is a good endpoint to study with a fair chance at reducing cases if environmental causes can be established. Hopefully, this research will be complemented by advances in treatments as well.

## References

Atlas, R.M., and Hazen, T.C. (2011). Oil biodegradation and bioremediation: A tale of the two worst spills in U.S. History. *Environ. Sci. Technol.* 45(16), 6709–6715. Doi: 10.1021/es2013227.

Bidleman, T.F., Jantunen, L.M., Hung, H., Ma, J., Stern, G.A., Rosenberg, B., and Racine, J. (2015). Annual cycles of organochlorine pesticide enantiomers in Arctic air suggest changing sources and pathways. *Atmos. Chem. Phys.* 15(3), 1411–1420. Doi: 10.5194/acp-15-1411-2015.

Butcher, L. (2020). A public policy agenda and movement to reduce Parkinson's disease burden gains momentum. *Neurol. Today* 20(20), 1–30. https://journals.lww.com/neurotodayonline/fulltext/2020/10220/a_public_policy_agenda_and_movement_to_reduce.6.aspx.

CARB. (2017). *California Air Toxics Program-Background.* California Air Resources Board. http://www.arb.ca.gov/toxics/background.htm (Accessed March 18, 2021).

CCOHS. (1997). *How Do Particulates Enter the Respiratory System.* Canadian Center for Occupational Health and Safety. http://www.ccohs.ca/oshanswers/chemicals/how_do.html accessed 3/18/21.

Chen, H., and Ritz, B. (2018). The search for environmental causes of Parkinson's disease: Moving forward. *J. Park. Dis.* 8, S9–S17.

Cook, M.B., Trabert, B., and McGlynn, K.A. (2011). Organochlorine compounds and testicular dysgenesis syndrome: Human data [Article]. *Int. J. Androl.*, 34(4), e68–e85. Doi: 10.1111/j.1365–2605.2011.01171.x.

Dowling, K.C., and Seiber, J.N. (2002). Importance of respiratory exposure to pesticides among agricultural populations. *Int. J. Toxicol.* 21(5), 371–381. Doi: 10.1080/10915810290096612.

Evangelou, E., Ntritsos, G., Chondrogiorgi, M., Kavvoura, F.K., Hernández, A.F., Ntzani, E.E., and Tzoulaki, I. (2016). Exposure to pesticides and diabetes: A systematic review and meta-analysis. *Environ. Int.* 91, 60–68. Doi: 10.1016/j.envint.2016.02.013.

Farrington, J.W. (2013). Oil pollution in the marine environment I: Inputs, big spills, small spills, and dribbles. *Environ. Sci. Policy Sustain. Develop.* 55(6), 3–13. Doi: 10.1080/00139157.2013.843980.

Ghaffarianhoseini, A., AlWaer, H., Omrany, H., Ghaffarianhoseini, A., Alalouch, C., Clements-Croome, D., and Tookey, J. (2018). Sick building syndrome: Are we doing enough? *Architec. Sci. Rev.* 61, 99–121.

Gill, R.J., Ramos-Rodriguez, O., and Raine, N.E. (2012). Combined pesticide exposure severely affects individual- and colony-level traits in bees. *Nature*, 491(7422), 105–108. Doi: 10.1038/nature11585.

Griffith, C.M., Baig, N., and Seiber, J.N. (2015). Contamination from industrial toxicants. In *Handbook of food chemistry* (pp. 719–751). Berlin Heidelberg: Springer. Doi: 10.1007/978-3-642-36605-5_11.

Hao, Y., Li, Y., Han, X., Wang, T., Yang, R., Wang, P., et al. (2019). Air monitoring of polychlorinated biphenyls, polybrominated diphenyl ethers and organochlorine pesticides in West Antarctica during 2011–2017: Concentrations, temporal trends and potential sources. *Environ. Pollut.* 249, 381–389.

IARC. (2018). *Agents Classified by the IARC Monograph.* International Agency for Research on Cancer from https://monographs.iarc.fr/wp-content/uploads/2018/09/ClassificationsAlphaOrder.pdf (Accessed March 18, 2021).

Jiao, L., Lao, Q., Chen, L., Chen, F., Sun, X., and Zhao, M. (2018). Concentration and influence factors of organochlorine pesticides in atmospheric particles in a coastal island in Fujian, Southeast China. *Aerosol. Air Qual. Res.* 18, 2982–2996.

Khanjani, N., Hoving, J.L., Forbes, A.B., and Sim, M.R. (2007). Systematic review and meta-analysis of cyclodiene insecticides and breast cancer. *J. Environ. Sci. Health, Part C*, 25(1), 23–52. Doi: 10.1080/10590500701201711.

Kiljanek, T., Niewiadowska, A., Semeniuk, S., Gaweł, M., Borzęcka, M., and Posyniak, A. (2016). Multi-residue method for the determination of pesticides and pesticide metabolites in honeybees by liquid and gas chromatography coupled with tandem mass spectrometry—Honeybee poisoning incidents. *J. Chromatogr. A*, 1435, 100–114. Doi: 10.1016/j.chroma.2016.01.045.

Kjuus, H. (2001). Acrylamide in tunnel construction - new (or old) lessons to be learned? *Scand. J. Work, Environ. Health*, 27(4), 217–218. Doi: 10.5271/sjweh.607.

Kovner, K. (2002). *Persistent Organic Pollutants: A Global Issue, A Global Response.* United States Environmental Protection Agency. Retrieved 2/10/2015 from http://www2.epa.gov/international-cooperation/persistent-organic-pollutants-global-issue-global-response%23table.

Laumbach, R.J., and Kipen, H.M. (2012). Respiratory health effects of air pollution: Update on biomass smoke and traffic pollution. *J. Allergy Clin. Immunol.* 129(1), 12–13. Doi: 10.1016/j.jaci.2011.11.021.

Lim, J.-e., Park, S.H., Jee, S.H., and Park, H. (2015). Body concentrations of persistent organic pollutants and prostate cancer: A meta-analysis. *Environ. Sci. Pollut. Res. Int.* 22(15), 11275–11284. Doi: 10.1007/s11356-015-4315-z.

Lopez-Avila, V., Wood, R., Flanagan, M., and Scott, R. (1987). Determination of volatile priority pollutants in water by purge and trap and capillary column gas chromatography/mass spectrometry. *J. Chromatog. Sci.* 25(7), 286–291. Doi: 10.1093/chromsci/25.7.286.

Luo, D., Zhou, T., Tao, Y., Feng, Y., Shen, X., and Mei, S. (2016). Exposure to organochlorine pesticides and non-Hodgkin lymphoma: a meta-analysis of observational studies. *Sci. Rep. (Nature Publisher Group).* 6, 25768. Doi: 10.1038/srep25768.

Mast, T.J., Hsieh, D.P.H., and Seiber, J.N. (1984). Mutagenicity and chemical characterization of organic constituents in rice straw smoke particulate matter. *Environ. Sci. Technol.* 18(5), 338–348. Doi: 10.1021/es00123a010.

Nadal, M., Marquès, M., Mari, M., and Domingo, J. (2015). Climate change and environmental concentrations of POPs: A review. *Environ. Res.* 143, 177–185. DOI: 10.1016/j.envres.2015.10.012.

NCEI NOAA. (2018a). *Water Temperature Table of the Alaska Coast.* National Centers for Environmental Information, National Oceanic and Atmospheric Administration. https://www.nodc.noaa.gov/dsdt/cwtg/alaska.html (Accessed March 18, 2021).

NCEI NOAA. (2018b). *Water Temperature Table of the Eastern Gulf of Mexico.* National Centers for Environmental Information, National Oceanic and Atmospheric Administration. https://www.nodc.noaa.gov/dsdt/cwtg/egof.html (Accessed March 18, 2021).

Ngo, M.A., Pinkerton, K.E., Freeland, S., Geller, M., Ham, W., Cliff, S., et al. (2010). Airborne particles in the San Joaquin Valley may affect human health. *California Agri.* 64(1), 12–16. Doi: 10.3733/ca.v064n01p12.

Oldroyd, B. P. (2007). What's killing American honey bees? *PLoS Biol.* 5(6), e168. Doi: 10.1371/journal.pbio.0050168.

San Joaquin Valley Air Pollution Control District. (2012). *Scientific Foundation and PM 2.5 Modeling Results.* http://www.valleyair.org/Workshops/postings/2012/10-9-12PM25/04Chapter4SciFoundationandModelingResults.pdf (Accessed March 18, 2021).

Schuster, P.F., Schaefer, K.M., Aiken, G.R., Antweiler, R.C., Dewild, J.F., Gryziec, J.D., et al. (2018). Permafrost stores a globally significant amount of mercury. *Geophys. Res. Lett.* 45(3), 1463–1471. Doi: 10.1002/2017gl075571.

Seiber, J.N., Coats, J., Duke, S.O., and Gross, A.D. (2014). Biopesticides: State of the art and future opportunities. *J. Agri. Food Chem.* 62(48), 11613–11619. Doi: 10.1021/jf504252n.

Seiber, J.N., Madden, S.C., McChesney, M.M., and Winterlin, W.L. (1979). Toxaphene dissipation from treated cotton field environments: component residual behavior on leaves and in air, soil, and sediments determined by capillary gas chromatography. *J. Agri. Food Chem.* 27(2), 284–291. Doi: 10.1021/jf60222a019.

Sun, Y., Hou, J., Cheng, R., Sheng, Y., Zhang, X., and Sundell, J. (2019). Indoor air quality, ventilation and their associations with sick building syndrome in Chinese homes. *Energy Build.* 197, 112–119.

Tang-Péronard, J.L., Andersen, H.R., Jensen, T.K., and Heitmann, B.L. (2011). Endocrine-disrupting chemicals and obesity development in humans: A review *Obesity Rev.* 12(8), 622–636. Doi: 10.1111/j.1467-789X.2011.00871.x.

Tareke, E., Rydberg, P., Karlsson, P., Eriksson, S., and Törnqvist, M. (2002). Analysis of acrylamide, a carcinogen formed in heated foodstuffs. *J. Agri. Food Chem.* 50(17), 4998–5006. Doi: 10.1021/jf020302f.

Taylor, A., and Birkett, J.W. (2020). Pesticides in cannabis: A review of analytical and toxicological considerations. *Drug Test. Analy.* 12, 180–190.

US EPA. (2015a). *Deepwater Horizon–BP Gulf of Mexico Oil Spill*. United States Environmental Protection Agency. http://www2.epa.gov/enforcement/deep water-horizon-bp-gulf-mexico-oil-spill (Accessed March 18, 2021).

US EPA. (2011). *Exposure Factors Handbook - Chapter 6: Inhalation Rates. 96. (Exposure factors handbook from US EPA*, Chapter 6, https://www.epa.gov/sites/production/ files/2015-09/documents/efh-chapter06.pdf.

US EPA. (2015b). *Information for Parents about Indoor Airquality*. United States Environmental Protection Agency. https://www.epa.gov/childcare/information-parents-about-indoor-air-quality. (Accessed March 18, 2021).

US EPA. (2016). *The Original List of Hazardous Air Pollutants*. United States Environmental Protection Agency. https://www3.epa.gov/airtoxics/188polls.html (Accessed March 18, 2021).

US EPA. (2021). https://www.epa.gov/criteria-air-pollutants#self (Accessed April 11, 2021).

vanEngelsdorp, D., Evans, J.D., Saegerman, C., Mullin, C., Haubruge, E., Nguyen, B.K., et al. (2009). Colony collapse disorder: A descriptive study (epidemiological survey of CCD). *PLoS One* 4(8), e6481. Doi: 10.1371/journal.pone.0006481.

Varma, R., and Varma, D. (2005). The Bhopal disaster of 1984. *Bullet. Sci. Technol. Soci.* 25, 37–45. Doi: 10.1177/0270467604273822.

Wang, Q., Wang, L., Li, X., Xin, J., Liu, Z., Sun, Y., et al. (2020). Emission characteristics of size distribution, chemical composition and light absorption of particles from field-scale crop residue burning in Northeast China. *Sci. Total Environ.* 710, 136304.

Wang, S., Salamova, A., Hites, R.A., and Venier, M. (2018). Spatial and seasonal distributions of current use pesticides (CUPs) in the atmospheric particulate phase in the Great Lakes region. *Environ. Sci. Technol.* 52(11), 6177–6186. Doi: 10.1021/acs. est.8b00123.

Wang, X., Banks, A.P.W., He, C., Drage, D.S., Gallen, C.L., Li, Y., Li, Q., Thai, P.K., and Mueller, J.F. (2019). Polycyclic aromatic hydrocarbons, polychlorinated biphenyls and legacy and current pesticides in indoor environment in Australia – occurrence, sources and exposure risks. *Sci. Total Environ* 693, 133588.

Wang, X., Sun, D., and Yao, T. (2016). Climate change and global cycling of persistent organic pollutants: A critical review. *Sci. China Earth Sci.* 59(10), 1899–1911. Doi: 10.1007/s11430-016-5073-0.

Wargocki, P. and Wyon, D.P. (2006). Research report on effects of HVAC on student performance. *ASHRAE J.* 48(10): 22–28.

Wargocki, P., Wyon, D.P., Sundell, J., Clausen, G., and Fanger, P.O. (2000). The effects of outdoor air supply rate in an office on perceived air quality, sick building syndrome (SBS) symptoms and productivity [Article]. *Indoor Air* 10(4), 222. Doi: 10.1034/j.1600-0668.2000.010004222.x.

Williams, R.L., Aston, L.S., and Krieger, R.I. (2004). Perspiration increased human pesticide absorption following surface contact during an indoor scripted activity program. *J. Expos. Analy. Environ. Epidemiol.* 14(2), 129–136. Doi: 10.1038/sj. jea.7500301.

Wilson, B., Hooper, M., Littrell, E., Detrich, P., Hansen, M., Weisskopf, C., and Seiber, J. (1991). Orchard dormant sprays and exposure of red-tailed hawks to organophosphates. *Bullet. Environ. Contaminat. Toxicol.* 47(5), 717–724. Doi: 10.1007/BF01701140.

Woodrow, J.E., Gibson, K.A., and Seiber, J.N. (2018). Pesticides and related toxicants in the atmosphere. In P. de Voogt, (Ed.), *Reviews of Environmental Contamination and Toxicology* (Vol. 247, pp. 147–196). Cham: Springer International Publishing.

Woodrow, J.E., Seiber, J.S., and Kim, Y.-H. (1986). Measured and calculated evaporation losses of two petroleum hydrocarbon herbicide mixtures under laboratory and field conditions. *Environ. Sci. Technol.* 20(8), 783–789.

Woodrow, J.E., Wong, J.M., and Seiber, J.N. (1989). Pesticide residues in spray aircraft tank rinses and aircraft exterior washes. *Bullet. Environ. Contaminat. Toxicol.* 42(1), 22–29.

Wu, X., Lam, J.C.W., Xia, C., Kang, H., Xie, Z., and Lam, P.K.S. (2011). Atmospheric concentrations of DDTs and chlordanes measured from Shanghai, China to the Arctic Ocean during the Third China Arctic research expedition in 2008. *Atmos. Environ.* 45(22), 3750–3757. Doi: 10.1016/j.atmosenv.2011.04.012.

Wu, X., Nethery, R.C., Sabath, M.B., Braun, D., and Dominici, F. (2020). Air pollution and COVID-19 mortality in the United States: Strengths and limitations of an ecological regression analysis. *Sci. Adv.* 6, eabd4049.

Xu, H., Du, S., Cui, Z., Zhang, H., Fan, G., and Yin, Y. (2011). Size distribution and seasonal variations of particle-associated organochlorine pesticides in Jinan, China. *J. Environ. Monitor.* 13(9), 2605. Doi: 10.1039/c1em10394f.

# 5

# Environmental Fate Models, with Emphasis on Those Applicable to Air

## 5.1 Introduction

In the past, not much thought was given to the fate of a chemical when it was released to the environment, as long as there was a perceived benefit (e.g., crop and residential pest control and the refinement of crude oil to provide products for industry and public use). However, from about the mid-point of the 20th century, there was a growing awareness that everything we humans do will impact the Earth in some way, often to its and our detriment. For example, releases of DDT, PCBs, and dioxins to the environment led to disastrous, nearly irreversible results. In addition, it was found that these chemicals, and many others, undergo long-range transport to become widely distributed in the environment. They persist and biomagnify in food chains as well as leave residues in soil, water, and biota (seals, eagles, polar bears—top-of-the-food-chain organisms, including humans). As a result of a growing awareness that the Earth's ecosystems are fragile, a grassroots environmental movement, sparked in part by Rachel Carson's *Silent Spring* (1962) (Carson, 1962), eventually led to the establishment of the U.S. Environmental Protection Agency (EPA) in 1970.

A more proactive approach to lessening chemical impact on the environment would be to predict the behavior and fate of a chemical before its use. With regard to this, the establishment in 1976 of the U.S. EPA Toxic Substances Control Act (TSCA), which requires extensive testing of substances before they are released, represented a move in the right direction. Testing might involve the use of empirical models in laboratory/field chambers that are spiked with the chemical(s) of interest. Historically, laboratory/field chambers, along with direct measurements in the open environment, generated the data that eventually were written into computer code. We now have numeric computer-based models as an alternative to the strictly empirical approach.

Modeling a chemical's potential behavior would go far in providing the data needed to make a judgment regarding the possible risk to humans and the environment. The modeling approach—chambers and computers—also

DOI: 10.1201/9781003217602-5

allows one to evaluate multiple sources of exposure and multiple chemicals simultaneously for risk assessment and risk management. Modeling is important for assessing environments with changing geographies, meteorology, and use (e.g., schools, hospitals, playgrounds), and wherever sensitive subpopulation exposures might occur. However, it is important to remember: "A model is an imitation of reality, which stresses those aspects that are assumed to be important, and omits all properties considered to be nonessential" (Schwarzenbach et al., 2002).

## 5.2 Empirical vs Computational Models

Empirical laboratory/field models can be divided into microcosms and mesocosms. A microcosm is typically enclosed in a laboratory chamber system (glass, Teflon®) that is constructed to simulate natural systems on a reduced scale, for measuring responses to varying conditions (e.g., moisture, nutrients, sunlight, and temperature) over time. When a chemical, or a mixture, is placed in the chamber, one can measure the water, air, plant, and animal concentrations of radio labelled precursors under different sets of conditions to obtain approximations of how the chemical might behave in the outdoor environment. A mesocosm is any outdoor experimental system that examines the natural environment under controlled conditions. Possible scenarios could include dosing experiments to evaluate the impact of chemical exposures (e.g., pesticides and solvents) on organisms or communities in their natural habitats. Mesocosm studies may be conducted in an enclosure or partial enclosure that is small enough so that key variables can be brought under control.

Numeric computer-based methods are attractive alternatives or ancillaries, ideally suited to the needs of the desk-bound scientist or regulator who does not have ready access to lab or field facilities. Numbers or data that represent environmental compartment variables (e.g., wind speed, temperature, sunlight intensity, and chemical properties) can be used as input to calculate the effect on the output. That is, it is possible to get immediate information on the effect of each variable. Volatilization, dispersal, and downwind transport models are useful in predicting exposure to downwind residents and workers, and the time needed for chemicals to be reduced in concentration to safe levels by ventilation or breakdown to hopefully by-products that are less toxic. Soil and water models are useful for predicting, for example, the movement of toxic chemicals into the groundwater or in runoff water. The processes are complex but can be described and predicted by modeling tools that have been relatively recently developed. Chemicals that are stable in the air can undergo long-range transport, a characteristic that can be quantified and used to encourage replacement or even banning of some chemicals.

With numeric computer-based models, the time course of the amount or concentration of the chemical in question in each compartment and final conditions at equilibrium constitute output, from which the rate of loss or dissipation of the chemical from each compartment can be calculated. This is a huge advantage for the regulator or emergency responder faced with a spill or other accident requiring decisions on how best to protect humans and wildlife from both immediate and long-term harm. You can play "what if" scenarios, such as the use of dispersants in an oil spill into water, or how extensive the effect might be on fish or other wildlife downwind or downstream from a spill. The extent and immediacy of evacuation of workers or nearby residents can be calculated; or, in the case of pesticide applications, the setting of buffer zones.

This chapter summarizes some examples of environmental fate models (empirical and computer-based). Included are CHEMEST and EPI Suite™, which are examples of physiochemical property calculation/estimation models for generating property data as input to the environmental fate models. Some models broadly address multimedia aspects of behavior and fate such as EPA's EXAMS or CalEPA's CalTOX, models used widely in regulatory risk assessments and other predictive efforts. Other models predict for just one or a few of the media involved in transport and fate, such as AQUATOX (water) or PRZM (plant soil root zone) or AERMOD and CALPUFF for air. Some are designed for specific classes of chemicals, such as pesticides, or fumigants (FEMS and PERFUM), or petroleum constituents. Examples of selected models follow.

## 5.3 Empirical Models

### 5.3.1 Empirical Models (Microcosm)

A microcosm can be any laboratory-contained model ecosystem (Table 5.1). It can contain environmental material (e.g., soil, water) brought into the lab for study purposes, or it can be a construct containing materials representative of an ecosystem (e.g., soil, water, plant, animal). The examples that follow are just a few representatives of the myriad studies to be found in the literature. Many of the microcosm studies published in recent years have focused on the use of microbes for the remediation of chemical pollutants, as illustrated below.

**TABLE 5.1**

Empirical Models

| Type | Category | Reference |
|------|----------|-----------|
| Chamber | Microcosm | www.epa.gov/sites/production/files/2015-07/documents/850-1900.pdf |
| Outdoor | Mesocosm | www.epa.gov/sites/production/files/2015-07/documents/850-1950.pdf |

Chambers for microcosm studies can vary in design (i.e., open or closed, shape, size) depending on the research needs/goals. For example, studies to monitor the microbial dechlorination of vinyl chloride, a carcinogen, in contaminated groundwater have been performed in small 160 mL serum bottles with vinyl chloride introduced to the headspace and monitored for the non-chlorinated degradation products (Findlay et al., 2016). By contrast, studies of aerobic microbial cometabolic degradation of trichloroethylene with toluene were performed in semicontinuous slurry microcosms that contained soil and water from a contaminated site (Han et al., 2007). Finally, studies to monitor the microbial degradation of petroleum hydrocarbons in contaminated soil have been performed in soil-filled columns through which air was passed to promote oxidation of the hydrocarbons by the microbes (Schaffner et al., 1998). The results of these microcosm studies were used to develop remediation strategies for contaminated ecosystems (see, for example, work done by GZA GeoEnvironmental, Inc.).

A chamber model that has been used for pesticide environmental fate studies is Metcalf's aquatic/terrestrial farm pond microcosm (Metcalf et al., 1971). This model was generally composed of a glass aquarium containing sand, water, and plants. It has been referred to as "an Illinois farm pond in a box," a quote attributed to Metcalf. The sand–water system contained snails, water fleas, algae, and plankton. After the system reached equilibrium, the plants were dosed with a solution containing a radiolabeled chemical. This construct allowed him to document meticulously, precisely, and with repeatability the movement of pesticides through the trophic web. Metcalf and his coworkers evaluated over 200 chemicals in this terrestrial–aquatic ecosystem, providing invaluable information on the environmental suitability not only of pesticides but also of industrial chemicals such as polychlorinated biphenyls and animal supplements. Information obtained from this model ecosystem, and others like it, confirmed information obtained laboriously from decades of field studies, validating the method as a fast, inexpensive, and reliable index of environmental fate.

Another study involving a pesticide was the assessment of the sensitivity of freshwater organisms—invertebrates and algae—to the fungicide fluazinam in single-species laboratory tests and together in microcosms (Van Wijngaarden et al., 2010). The study showed that a measure of species sensitivity was the same for the microcosm communities as in the single-species tests, showing the utility of the microcosm approach.

## 5.3.2 Empirical Models (Mesocosm)

Mesocosms are somewhat in-between "real-world conditions" and the artificial confines of a laboratory chamber, but with less control over variables (Table 5.1). Mesocosms can vary in size and complexity, depending on the research needs. One document describes a mesocosm design for pesticide registration efforts (Touart, 1980). The design calls for a pond of about 0.1 acres in surface area containing at least $300\,m^3$ water at a maximum depth of $2\,m$. The design also prescribes the type and depth of the bottom sediment and "representative" biota, including fish. A minimum of 12 ponds is required for statistical analysis. Recommended pesticide dosing levels are 0.1× and 10× the intermediate level, which is determined based on modeling and experiential data. Another study looked at the possible effect of mesocosm pool size on algal growth (especially the effect of change in surface area/volume ratio) and found little, if any, effect. The conclusion was that study results using relatively small mesocosms could be extrapolated to larger artificial and natural systems (Spivak et al., 2011). By contrast, a study showed the difficulty of extrapolating marine mesocosm results to the open ocean for computer model validation (Watts and Bigg, 2001). A final example is a study using a mesocosm to monitor the impact of a marine pathogen on a fish (sharpsnout seabream) raised in an aquaculture environment (Katharios et al., 2015). The authors concluded: "These results demonstrate the advantage of mesocosm studies for investigating the effect of environmental bacteria on susceptible hosts and provide an important insight into the genome dynamics of a novel fish pathogen."

## 5.4 Numeric Models (Computer)

Chamber models, such as those discussed above, are typically used in a lab/greenhouse/field plot setting. They allow control of some variables and, in some cases, can incorporate radiolabeled compounds. They are good for ranking chemicals in terms of such parameters as biodegradability and ecological magnification. They are also good for understanding how structures and key physicochemical properties influence behavior and fate. The data set accumulated over the years from using these model systems has been converted into computer code for relatively rapid estimation of chemical environmental fate using numeric models. Although they are time-consuming and labor-intensive, chamber models are still used as needed (e.g., to validate a computer model).

Numeric models allow one to simulate environmental behavior using only a computer. They give answers relatively quickly, which is important when a real-life escape or spill of a chemical occurs. And if it becomes necessary to implement evacuation plans or make other decisions requiring actions. They also allow one to do "what if" assessments—what if the chemical had a non-polar substituent added, or had its vapor pressure reduced, or the daytime high temperature during a spray application was 35°C vs 15°C, etc.? There are many types of models that we deal with in daily life: economic, weather, traffic congestion, and fire danger. These all allow us to paint a realistic scenario for the future—not necessarily exact, but close enough to base some decisions on.

### 5.4.1 Water/Multimedia Models

EXAMS (exposure assessment modeling system) is an EPA-developed multimedia, equilibrium computer model that was first published in 1982 (Burns et al., 1982) (Table 5.2). It provides interactive computer software for formulating aquatic ecosystem models and rapidly evaluating the fate, transport, and exposure concentrations of various synthetic organic chemicals—pesticides, industrial materials, and leachates from disposal sites. EXAMS provides facilities for long-term (steady-state) analysis of chronic chemical discharges, initial-value approaches for study of short-term chemical releases, and full kinetic simulations that allow for monthly variation in mean climatological parameters and alteration of chemical loadings on daily time scales (Burns et al., 1982). The equilibrium output gives practical information that can be used to direct sampling efforts, or determine which organisms (aquatic, terrestrial) might be expected to have the greatest exposure. In one application, EXAMS was successfully used to model volatilization losses of two pesticides (molinate and methyl parathion) from a flooded rice field (Seiber et al., 1989).

AQUATOX is another multimedia simulation model for aquatic systems (Table 5.2). The model predicts the fate of various pollutants, such as nutrients

**TABLE 5.2**

Water Quality Models

| Name | Description | Reference |
|------|-------------|-----------|
| EXAMS (Exposure Analysis Modeling System) | An interactive software for formulating aquatic ecosystem models and evaluating fate, transport, and exposure of pesticides, industrial materials, and leachates from disposal sites. | http://www2.epa.gov/exposure-assessment-models/exams-version-index |
| Positive matrix factorization (PMF) | A mathematical receptor model developed by the EPA for the development and review of air and water quality standards, exposure research, and environmental forensics. Can analyze: sediments, wet deposition, surface water, ambient air, and indoor air. | http://www2.epa.gov/air-research/positive-matrix-factorization-model-environmental-data-analyses |
| AQUATOX | Simulation model that predicts the fate of various pollutants, nutrients, organic chemicals, and their effects on ecosystems including fish, invertebrates, and aquatic plants | http://www2.epa.gov/exposure-assessment-models/aquatox |
| Hydrodynamic, sediment, and contaminant transport model (HSCTM2D) | Simulates two-dimensional, vertically integrated, surface water flow, sediment transport, and contaminant transport | http://www2.epa.gov/exposure-assessment-models/hsctm2d |

and organic chemicals, and their effects on the ecosystem, including fish, invertebrates, and aquatic plants. It is a valuable tool for ecologists, biologists, water quality modelers, and anyone involved in performing ecological risk assessments for aquatic ecosystems. From the beginning, AQUATOX was developed as an applied model for use by environmental analysts. It incorporates constructs from classic ecosystem and chemodynamic models. A private consulting firm, under contract with the EPA, performed an extensive search of the environmental literature and identified 173 articles describing the use of AQUATOX (EPA, 2013). AQUATOX holds the most potential for generalized risk assessment use and also has potential for use in regulatory scenarios including development of water quality criteria, total maximum daily loads, and analysis of management alternatives. One detailed AQUATOX case study was concerned with the effects of a pesticide (dieldrin) on largemouth bass in an Iowa reservoir (Mauriello and Park, 2002). The objective of maintaining a viable recreational bass fishery was used to evaluate alternative strategies for reducing risks to the bass population. The model quantified the potential effectiveness of alternative management strategies and provided feedback for policy refinements.

Another multimedia modeling approach is the fugacity concept, developed by Mackay and coworkers (Table 5.3). It is fundamental to current environmental modeling where fugacity is defined as the "escaping tendency" of a chemical substance from a phase in terms of mass transfer coefficients. The concept is described in the book *Multimedia Environmental Models: The Fugacity Approach* (Mackay, 2001) and many applications are discussed in the book *Handbook of Chemical Mass Transport in the Environment* (Thibodeaux and Mackay, 2010). Some specific examples of the use of the fugacity concept include predicting the environmental fate of multiple chemicals (Cahill et al., 2003) and modeling pharmaceutical residues in sewage (Khan and Ongerth, 2004). The Cahill et al. study used the framework of the Mackay fugacity model, but with a few key changes: (1) the model handled up to four independent chemical species which allowed for the analyses of the parent chemical and degradation products; (2) the model is dynamic instead of steady-state and can predict chemical concentrations in each medium as a function of time, so it can measure the dissipation of a chemical; and (3) the evaluative

**TABLE 5.3**

Fugacity Models

| Name | Description | Reference |
| --- | --- | --- |
| Fugacity Models I-III | Distribution-based models that incorporate all environmental compartments and are based on the steady-state fluxes of pollutants across compartment interfaces. Other fugacity models include the multimedia fugacity model and multimedia equilibrium criterion model | http://www.kch.tul.cz/ sedlbauer/fugacity_ model.pdf |

environment of the model has been modified to better describe the behavior of highly water-soluble chemicals (Cahill et al., 2003). An additional feature is the ability to simulate cycling and branching in the degradation pathways. This article used chlorpyrifos, pentachlorophenol, and perfluorooctane sulfonate to demonstrate the capabilities of this new model (Cahill et al., 2003). Khan and Ongerth made a theoretical model using the fugacity concept that provides predictions of pharmaceutical concentrations, behaviors, and fates in raw, primary, and secondary sewage (Khan and Ongerth, 2004). The model incorporates two main features: (1) the prediction of pharmaceutical concentrations at the sewage treatment plant inlet, and (2) the prediction of pharmaceutical concentrations, behavior and fate during primary and secondary sewage treatment. Although the model currently has a relatively "high degree of uncertainty," it does provide (1) a basis for estimating the relationship between the quantity of pharmaceuticals used and the observed concentrations of the measured compounds; (2) an estimate of concentration of compounds that have not been measured; and (3) an estimation of future concentrations of new drugs.

### 5.4.2 Soil/Root Zone Models

PRZM (pesticide root zone model) is a one-dimensional, finite-difference model that accounts for pesticide and nitrogen fate in the crop root zone (Table 5.4). PRZM3 includes modeling capabilities for such phenomena as soil temperature simulation, volatilization, and vapor-phase transport in soils, irrigation simulation, microbial transformation, and a method of characteristics algorithm to eliminate numerical dispersion. PRZM is capable of simulating transport and transformation of the parent compound and as many as two daughter species. VADOFT is a one-dimensional, finite-element code that solves the Richard's equation for flow in the unsaturated zone. The user may make use of constitutive relationships between pressure, water content, and hydraulic conductivity to solve the flow equations. VADOFT may also simulate the fate of two parent and two daughter products. The PRZM and VADOFT codes are linked together with the aid of a flexible execution supervisor that allows the user to build loading models that are tailored to site-specific situations. In order to perform probability-based exposure assessments, the code is also equipped with a Monte Carlo pre- and post-processor (www.epa.gov/exposure-assessment-models/przm-version-index).

Root Zone Water Quality Model 2 (RZWQM2) simulates major physical, chemical, and biological processes in an agricultural crop production system (Table 5.4). RZWQM2 is a one-dimensional (vertical in the soil profile) process-based model that simulates the growth of the plant and the movement of water, nutrients, and pesticides over, within, and below the crop root zone of a unit area. It has a quasi-two-dimensional macropore/lateral flow. It responds to agricultural management practices including planting and harvest practices, tillage, pesticide, manure and chemical nutrient applications,

**TABLE 5.4**

Pesticide/Soil Models

| Name | Description | Reference |
|------|-------------|-----------|
| CalTOX | A multimedia total exposure model for hazardous-waste sites | https://dtsc.ca.gov/caltox/ |
| PRZM Version Index | Links two subordinate models, PRZM and VADOFT, in order to predict pesticide transport and transformation down through the crop root and unsaturated zone. Can model: soil temperature simulation, volatilization, and vapor phase transport in soils, irrigation simulation, microbial transformation, and a method of characteristics (MOC) algorithm to eliminate numerical dispersion | http://www2.epa.gov/exposure-assessment-models/przm-version-index |
| RZWQM (root zone water quality model) | Simulates major physical, chemical, and biological processes in an agricultural crop production system | www.ars.usda.gov/Research/docs.htm?docid=17740 |

and irrigation events. The model includes simulation of a tile drainage system. It has a Windows Interface (RZWQM2.EXE), which manages input and output for Projects and Scenarios and executes the science model (RZWQMrelease. exe). The science may also execute off ASCII file IO. RZWQM2 may be used as a tool for assessing the productivity of various cropping systems for various soil, weather, and management conditions. Once calibrated and validated to the productivity of a cropping system for a climatic region, alternate soils and crop management scenarios may be tested for development of best management practices for the region with regard to crop productivity and environmental sustainability. Testing of these managements through historical climates can provide production probability distribution functions based on past climate patterns. Monthly weather modifiers are provided in RZWQM2 to test cropping system responses to increase or decrease in factors such as temperature, radiation, wind, relative humidity, and $CO_2$. Management modifiers allow users to game with application amounts. It is currently being tested for its adequacy to implement effects of climate change on systems (www.ars. usda.gov/Research/docs.htm?docid=17740).

These two models and one more—OpusCZ—were evaluated for their accuracy in simulating pesticide runoff at the edge of agricultural fields (Zhang and Goh, 2015). The investigators found that, for runoff generated by sprinkler irrigation and rainfall, all the models were equally accurate. However, for runoff generated by flood irrigation, RZWQM and OpusCZ were more accurate.

## 5.4.3 Air/Dispersion and Fate Models

Many computer-based environmental models for the atmosphere have been developed to describe the behavior and fate of a chemical under almost every imaginable set of conditions: point sources, area sources, multiple sources, variable emission rates, downwind deposition rates, simple terrain, complex terrain, building downwash, variable temperature, variable windspeed, etc. (Table 5.5). The goal of many air dispersion models is to estimate chemical concentrations at different distances downwind of a source, or concentrations in an air basin from multiple sources (e.g., AERMOD; ISCST [Industrial Source Complex-Short Term]). In addition, some dispersion models (e.g., CALPUFF) can estimate chemical conversion rates to calculate concentrations of breakdown products, some of which might be more toxic than the parent (e.g., parathion to paraoxon, methyl isothiocyanate (MITC) to methyl isocyanate (MIC)).

A popular software suite that is used for monitoring accidental releases of toxic chemicals is CAMEO (computer-aided management of emergency operations [ALOHA; MARPLOT]) (Table 5.5). It can be used by an emergency

## TABLE 5.5

Air/Pollutant Models

| Name | Description | Reference |
|------|-------------|-----------|
| AERMOD modeling system | A steady-state plume model that incorporates air dispersion based on planetary boundary layer turbulence structure and scaling concepts | https://www.epa.gov/scram/air-quality-dispersion-modeling-preferred-and-recommended-models |
| CALPUFF | An advanced, integrated Lagrangian puff modeling system for the simulation of atmospheric pollution dispersion | http://src.com/calpuff/calpuff1.htm |
| Fused air quality surfaces using downscaler model | Combines daily ozone and particulate matter monitoring and modeling data from across the United States to provide improved fine-scale estimates of air quality in communities and other specific local regions | http://www2.epa.gov/air-research/-fused-air-quality-surfaces-using-downscaling-tool-predicting-daily-air-pollution |
| PMF | A mathematical receptor model developed by the EPA for the development and review of air and water quality standards, exposure research, and environmental forensics. Can analyze sediments, wet deposition, surface water, ambient air, and indoor air | http://www2.epa.gov/air-research/-positive-matrix-factorization-model-environmental-data-analyses |
| Community multiscale air quality (CMAQ) model | Computational tool that simultaneously models multiple air pollutants including ozone, particulate matter, and air toxics to help air quality managers determine scenarios. Provides detailed information on air pollutant concentrations in any given area for any specified emission or climate scenario | http://www2.epa.gov/air-research/-community-multi-scale-air-quality-cmaq-modeling-system-air-quality-management |

*(Continued)*

**TABLE 5.5 (*Continued*)**

Air/Pollutant Models

| Name | Description | Reference |
|------|-------------|-----------|
| Industrial Source Complex Short Term Model (ISCST3, also called ISC3) | A Gaussian plume model that assesses pollutant concentrations from a wide variety of sources associated with an industrial complex. Can account for the following: settling and dry deposition of particles; downwash; point, area, line, and volume sources; plume rise as a function of downwind distance; separation of point sources; and limited terrain adjustment | https://www.epa.gov/scram/air-quality-dispersion-modeling-alternative-models |
| ISC-Prime (Plume Rise Model Enhancements) | A model with building downwash incorporated into the ISCST3 model | https://www.epa.gov/scram/air-quality-dispersion-modeling-alternative-models |
| CAMEO® software suite | Computer-aided management of emergency operations; software suite for accidental spills/releases | http://www.epa.gov/cameo |
| Fumigant emission modeling system (FEMS) | Models airborne exposures to agricultural fumigants like pesticides. Calculates emission rates over time and incorporates meteorological parameters | http://sullivan-environmental.com/fems/ |
| Probabilistic exposure and risk model for fumigants (PERFUM) | Addresses bystander exposures to fumigants following agricultural applications. Developed to address the need for buffer zones for fumigants. Can model agricultural fields, and multiple fields emitting fumigants simultaneously | https://www.exponent.com/experience/-probablistic-exposure-and-risk-model-for-fumigants/ |
| Support center for regulatory air models (SCRAM) | Provides quantitative models to predict the dispersion of air pollution | http://www3.epa.gov/airquality/modeling.html |

response team en route to the spill/release site to determine if a nearby neighborhood needs to be evacuated, based on the model plot of the developing chemical plume. Most people use the CAMEO programs to respond to or plan for accidental chemical releases. However, some users have gone beyond the basic uses, for example: (1) aerial ambulance companies have used the MARPLOT part of CAMEO to provide the direction and distance to local hospitals to help expedite patient transport, and (2) first responders make use of the entire CAMEO software suite of programs at the weapons of mass destruction training developed by the Department of Homeland Security. After entering chemical inventories and special locations into CAMEO*fm*, some planners assess likely terrorist targets within their area. After Hurricanes Katrina and Rita, emergency responders used the CAMEO suite to complete challenging response tasks such as (1) estimating the number of affected residences in New Orleans, (2) mapping evacuation routes and collection sites for hazmat containers displaced by the storm, (3) defining exclusion zones around dangerous hazmat containers, and (4) selecting safety gear for workers handling hazardous debris. The United Nations Environment Programme (UNEP) selected the CAMEO suite as a tool to help developing nations prepare for—and respond to—chemical accidents. Under UNEP's Awareness and Preparedness for Emergencies at the Local Level (APELL) program, CAMEO has been demonstrated or taught in 50 countries.

Another popular air dispersion model is U.S. EPA's AERMOD (Table 5.5). The AMS/EPA Regulatory Model (AERMOD) is an air dispersion model based on planetary boundary layer theory. It is a steady-state dispersion model designed for short-range (up to 50 km) dispersion of air pollutant emissions from stationary industrial sources. AERMOD utilizes a similar input and output structure to ISCST3 and shares many of the same features, as well as offering additional features. It is used extensively to assess pollution concentration from a wide variety of sources. A few examples of the many uses of the dispersion model include an assessment of exposure to air pollutants ($NO_2$ and $SO_2$) in an industrial complex setting in Thailand (Jittra et al., 2015) and assessment of mass loss of sulfuryl fluoride from two structure fumigation operations in California (Tao, 2019). In the Thailand study, emission data were obtained from 292 point sources in an industrial area. Modeled concentration data were compared to measured data from 10 receptor sites. Overall results revealed that AERMOD provided accurate predictions compared to the measured concentrations for both $NO_2$ and $SO_2$. In the California study, total mass loss of the fumigant from the tarped structures was determined, based on measured air concentrations, and estimated using AERMOD with a back calculated flux value (Johnson et al., 2010; Ross et al., 1996). The estimated total mass loss compared well with the measured values (see also Chapter 8, Fumigants).

CALPUFF is an advanced, integrated Lagrangian puff modeling system for the simulation of atmospheric pollution dispersion (Table 5.5). Unlike steady-state Gaussian models such as AERMOD, CALPUFF allows variable/curve plume trajectories, variable meteorological conditions, accurate treatment of calm hours and low wind speed conditions, while retaining information of previous hours' emissions. Since CALPUFF is the preferred model for >50 km long-range transport, one case study used the model to predict the plume shape and concentrations of tracers (e.g., perfluorocarbons and $SF_6$) 100 and 600 km downwind of a release point (Irwin, 1998). Overall, CALPUFF-generated plume shapes compared well with what were measured using arcs of samplers. However, the centerline of the modeled plumes was offset from the measured plumes by as much as 17°. Modeled centerline concentrations at 100 km were 1.5–2.0 times greater than the measured concentrations (e.g., 1.05 ppt [actual] vs. 1.80 ppt [model]). At 600 km, the modeled centerline concentrations were less than the measured values, but both were the same order of magnitude (e.g., 0.38 ppt [actual] vs. 0.13 ppt [model]).

### 5.4.3.1 Emission Rate Estimation for Air/Dispersion and Fate Models

To give reliable results, dispersion models need, as input, accurate emission rates. Modeling the dispersion of air pollutants from industrial complexes relies on already known stack emission rates. However, in other situations, the emission rates are not readily known and must be determined before dispersion modeling can be done. For example, emission rates for pesticide-treated fields can be estimated using the numeric correlations derived from emission rate data measured in the field and laboratory (Woodrow et al., 1997, 2001, 2011). These investigators correlated the physicochemical properties (i.e., vapor pressure, water solubility, soil adsorption coefficient) of a series of volatile and semivolatile pesticides with measured emission rates (Woodrow et al., 1997, 2011). The physicochemical properties were taken from published literature values. However, in one study, these investigators estimated the vapor pressures for a series of semivolatile pesticides using techniques discussed in Section 5.5 (Woodrow et al., 2001). They used their correlations with a dispersion model (SCREEN3 [same algorithms as in ISCST3]) to estimate downwind air concentrations and found that their estimated values agreed well with the measured values. In another approach, the emission rate for a pesticide can be back calculated using a few downwind concentrations in a dispersion model (e.g., AERMOD, CALPUFF, ISCST3) (Johnson et al., 2010; Ross et al., 1996). The result can then be used to estimate total mass loss of pesticide from the treated field and concentrations further downwind for exposure assessments.

## 5.5 Chemical Property Estimation

To use these models and correlate the output with the physicochemical properties of the chemical, it is important to have a good set of physicochemical properties that go along with the chemical(s) of interest. Property data may be available in the literature, or more likely they will need to be calculated or estimated.

There are several potential sources of physicochemical properties:

- Literature search
- Google or Wikipedia Search
- Reference books (e.g., *Illustrated Handbook of Physical–Chemical Properties and Environmental Fate for Organic Chemicals*, Vol. V-Pesticide Chemicals (Mackay et al., 1997)).
- Estimation methods keyed to the structure of one or more physicochemical properties (e.g., *Chemical Property Estimation—Theory and*

**TABLE 5.6**

Chemical Property Estimation and Calculation Methods to Support Model Use

| Name | Description | Reference |
|------|-------------|-----------|
| CHEMEST (chemical property estimation) | Software to estimate chemical properties | Lyman, W. J., R. G. Potts, and G. C. Magil. (1983). *CHEMEST: User's Guide; a Program for Chemical Property Estimation.* Cambridge: AD Little Inc. |
| Estimation programs interface (EPI) Suite™ | Estimates: gas-phase reaction rates, Henry's law constants, melting point, boiling point, aerobic and anaerobic biodegradability, biodegradation half-life, organic carbon-normalized sorption coefficient, log octanol–water partition coefficient using atom/fragment contributions | http://www2.epa.gov/tsca-screening-tools/epi-suitetm-estimation-program-interface |
| GCSOLAR | Computes the photolysis rate and half-lives of pollutants in the aquatic environment | http://www2.epa.gov/exposure-assessment-models/gcsolar |

*Application* (Baum, 1997), *Handbook of Property Estimation Methods for Chemicals—Environmental and Health Sciences* (Mackay and Boethling, 2000), or *Handbook of Chemical Property Estimation Methods—Environmental Behaviour of Organic Compounds* (Lyman et al., 1990)).

- Experimental generation
- Software (Table 5.6)

### 5.5.1 EPI Suite™

EpiSuite™ (Estimation Programs Interface) is a Windows database of physico-chemical properties and environmental fate estimation programs developed by the EPA's Office of Pollution Prevention Toxics and the Syracuse Research Corporation (Table 5.6). Epi Suite™ uses a single input to run several estimation programs. It is readily accessible and allows one to rapidly obtain several properties, for one or several chemicals (see Table 4.1). A compound can be inputted with its name, CAS number, or simplified molecular input line entry system (SMILES) string. The SMILES string is a chemical shorthand characterization for a given structure. Figure 5.1 uses morphine as an example of a SMILES string. There is some consistency in the resulting values, which is important for comparing the behavior of several different chemicals. If one were to take original literature values for one chemical, these may not be consistent and comparable with values for other chemicals. Use of Epi Suite™ is thus advisable, at least for a first-cut estimation.

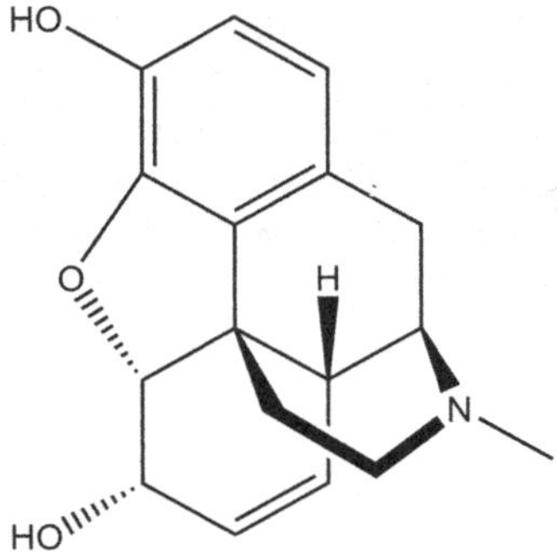

**FIGURE 5.1**
Structure of morphine. Example of EPI Suite™ SMILES notation for morphine: C1=CC2C
(N(C)C5)Cc3ccc(O)c4c3C2(C5)C(O4)C1O.

## 5.5.2 Literature Search

There are also a number of regression equations available in the literature for estimating a property via a correlation equation for one or more other properties. For example, there is a good correlation between $K_{ow}$ and water solubility, BCF, and $K_{oc}$. One can use the experimental $K_{ow}$ value, or the estimation of $K_{ow}$ via a Leo-Hansch calculation from the structure as a starting point.

## 5.5.3 Experimental Generation

Alternatively, one can estimate $K_{ow}$ and related polarity-based properties from high-pressure liquid chromatography (HPLC) retention data relative to a standard for which most properties are well known, such as di-n-butyl phthalate. Similarly, vapor pressure can be estimated from gas chromatography (GC) retention data relative to a standard like di-n-butyl phthalate or aldrin. It is advisable to estimate properties by more than one method and compare the estimated values with experimental values in the literature. This allows for a science-based judgment of the "best" values.

## 5.6 Conclusions

There have been substantial efforts to produce numeric computer-based models that can mimic the environmental behavior and fate of chemical pollutants, as an adjunct to, or replacement for experimental study of the pollutants under environmentally relevant conditions using microcosms/mesocosms. As shown in this chapter, such models can be useful to scientists, regulators, and the general population when faced with addressing questions related to the outcome of pollutant residues in the environment. In addition to those listed here, more extensive lists of models can be found at the following links: (1) air models—http://www.ehssoftserve.com/air_airmod.htm and (2) water models—water.usgs.gov/software/lists/groundwater, water.usgs.gov/software/lists/surface_water, and www.epa.gov/nscep.

## 5.7 Further Reading

Good references to the available environmental fate models and their uses can be found in the following books:

1. *Multimedia Environmental Models: The Fugacity Approach* by Mackay
2. *Handbook of Chemical Mass Transport in the Environment* (edited by Thibodeaux and Mackay)
3. *Modeling the Fate and Effect of Toxic Substances in the Environment* by Jorgensen
4. *Chemical Fate and Transport in the Environment* by Hemond and Fechner
5. *A Basic Introduction to Pollutant Fate and Transport: An Integrated Approach with Chemistry, Modeling, Risk Assessment, and Environmental Legislation* by Dunnivant and Anders
6. *Environmental Fate and Transport Analysis with Compartment Modeling* by Little and Francis

## References

Baum, E. (1997). *Chemical Property Estimation: Theory and Application*. Boca Raton: CRC Press.

Burns, L.A., Cline, D.M., and Lassiter, R.R. (1982). *Exposure Analysis Modeling System (EXAMS) User Manual and System Documentation*. Athens, GA: Environmental Protection Agency.

Cahill, T.M., Cousins, I., and Mackay, D. (2003). General fugacity-based model to predict the environmental fate of multiple chemical species. *Environ. Toxicol. Chem. Int. J.* 22, 483–493.

Carson, R. (1962). *Silent Spring*. Boston, MA: Houghton Mifflin.

EPA (2013). *Annotated Bibliography for AQUATOX*. Bethesda, MD: Environmental Protection Agency.

Findlay, M., Smoler, D.F., Fogel, S., and Mattes, T.E. (2016). Aerobic vinyl chloride metabolism in groundwater microcosms by methanotrophic and etheneotrophic bacteria. *Environ. Sci. Technol.* 50, 3617–3625.

Han, Y., Kuo, M.T., Tseng, I., and Lu, C. (2007). Semicontinuous microcosm study of aerobic cometabolism of trichloroethylene using toluene. *J. Hazard. Mater.* 148, 583–591.

Irwin, J.S. (1998). A Comparison of CALPUFF modeling results with 1977 INEL field data results. In *Air Pollution Modeling and Its Application XII.* (pp. 143–153). Boston, MA: Springer.

Jittra, N., Pinthong, N., and Thepanondh, S. (2015). Performance evaluation of AERMOD and CALPUFF air dispersion models in industrial complex area. *Air Soil Water Res.* 8, ASWR-S32781.

Johnson, M., Isakov, V., Touma, J., Mukerjee, S., and Özkaynak, H. (2010). Evaluation of land-use regression models used to predict air quality concentrations in an urban area. *Atmos. Environ.* 44, 3660–3668.

Katharios, P., Seth-Smith, H.M., Fehr, A., Mateos, J.M., Qi, W., Richter, D., Nufer, L., Ruetten, M., Soto, M.G., and Ziegler, U. (2015). Environmental marine pathogen isolation using mesocosm culture of sharpsnout seabream: striking genomic and morphological features of novel Endozoicomonas sp. *Sci. Rep.* 5, 1–13.

Khan, S.J., and Ongerth, J.E. (2004). Modelling of pharmaceutical residues in Australian sewage by quantities of use and fugacity calculations. *Chemosphere* 54, 355–367.

Lyman, W. J., R. G. Potts, and G. C. Magil. (1983). *CHEMEST: User's Guide; a Program for Chemical Property Estimation.* Cambridge: AD Little Inc.

Lyman, W.J., Reehl, W.F., and Rosenblatt, D.H. (1990). *Handbook of Chemical Property Estimation Methods.* Washington, D.C.: American Chemical Society.

Mackay, D. (2001). *Multimedia Environmental Models: The Fugacity Approach.* Boca Raton: CRC Press.

Mackay, D., and Boethling, R.S. (2000). *Handbook of Property Estimation Methods for Chemicals: Environmental Health Sciences* Boca Raton: CRC Press.

Mackay, D., Shiu, W.Y., and Ma, K.-C. (1997). *Illustrated Handbook of Physical-Chemical Properties and Environmental Fate for Organic Chemicals*, Volume V - Pesticide Chemicals. Boca Raton: CRC Press.

Mauriello, D.A., and Park, R.A. (2002). *An Adaptive Framework for Ecological Assessment and Management.* International Congress on Environmental Modelling and Software. 5. Lugano, Switzerland: Brigham Young University Scholars Archive.

Metcalf, R.L., Sangha, G.K., and Kapoor, I.P. (1971). Model ecosystem for the evaluation of pesticide biodegradability and ecological magnification. *Environ. Sci. Technol. 5*, 709–713.

Ross, L.J., Johnson, B., Kim, K., and Hsu, J. (1996). Prediction of methyl bromide flux from area sources using the ISCST model. *J. Environ. Q.* 25, 885–891.

Schaffner, I., Wieck, J.M., Wright, C.F., Katz, M., and Pickering, E. (1998). *Microbial Enumeration and Laboratory-Scale Microcosm Studies in Assessing Enhanced Bioremediation Potential of Petroleum Hydrocarbons. In the 11th Annual Conference on Contaminated Soils*, University Massachusetts at Amherst.

Schwarzenbach, R.P., Gschwend, P.M., and Imboden, D.M. (2002). *Environmental Organic Chemistry.* Hoboken, NJ: Wiley-Interscience, p. 947.

Seiber, J.N., McChesney, M.M., and Woodrow, J.E. (1989). Airborne residues resulting from use of methyl parathion, molinate and thiobencarb on rice in the Sacramento Valley, California. *Environ. Toxicol. Chem. Int. J.* 8, 577–588.

Spivak, A.C., Vanni, M.J., and Mette, E.M. (2011). Moving on up: can results from simple aquatic mesocosm experiments be applied across broad spatial scales? *Freshw. Biol.* 56, 279–291.

Tao, J. (2019). Estimating sulfuryl fluoride emissions during structural fumigation of residential houses. *Water. Air. Soil Pollut. 230*, 1–10.

Thibodeaux, L.J., and Mackay, D. (2010). *Handbook of Chemical Mass Transport in the Environment*. Boca Raton: CRC Press.

Touart, L. (1980). *Aquatic Mesocosm Tests to Support Pesticide Registrations*. Washington, D.C.: Environmental Protection Agency.

Van Wijngaarden, R., Arts, G., Belgers, J., Boonstra, H., Roessink, I., Schroer, A., and Brock, T. (2010). The species sensitivity distribution approach compared to a microcosm study: A case study with the fungicide fluazinam. *Ecotoxicol. Environ. Saf. 73*, 109–122.

Watts, M.C., and Bigg, G.R. (2001). Modelling and the monitoring of mesocosm experiments: two case studies. *J. Plankton Res. 23*, 1081–1093.

Woodrow, J.E., Seiber, J.N., and Baker, L.W. (1997). Correlation techniques for estimating pesticide volatilization flux and downwind concentrations. *Environ. Sci. Technol. 21*, 523–529.

Woodrow, J.E., Seiber, J.N., and Dary, C. (2001). Predicting pesticide emissions and downwind concentrations using correlations with estimated vapor pressures. *J. Agric. Food Chem. 49*, 3841–3846.

Woodrow, J.E., Seiber, J.N., and Miller, G.C. (2011). Correlation to estimate emission rates for soil-applied fumigants. *J. Agric. Food Chem. 59*, 939–943.

Zhang, X., and Goh, K.S. (2015). Evaluation of three models for simulating pesticide runoff from irrigated agricultural fields. *J. Environ. Qual. 44*, 1809–1820.

# 6

## Sampling and Analysis

### 6.1 Introduction

Pesticides are being applied in urban and agricultural settings at an annual rate of almost 3 billion kg worldwide and at greater than 500 million kg in the United States. Although these materials are applied to specific targets, such as soil, water, or plant foliage, pesticide residues can be unintentionally transported from the target site through the air (atmosphere). Often, half or more of applied pesticides are emitted to the air (Majewski et al., 1990). Once airborne, pesticides may move downwind, where they can affect nontarget organisms such as vegetation, aquatic and terrestrial wildlife, and humans. Nevertheless, these chemicals are vital for control of pests.

Assessment of nontarget impacts of pesticides requires that pesticide transport from the source region be accurately quantified. Applications have been developed by academic research laboratories and regulatory agencies (see the Glossary for descriptions of U.S. Environmental Protection Agency (EPA), Occupational Safety and Health Administration (OSHA), and United States Department of Agriculture (USDA) Animal and Plant Health Inspection Service (APHIS)) for most volatile and semivolatile pesticides and related contaminants including fumigants such as methyl bromide ethylene oxide/propylene oxide, 1,3-D, chloropicrin, and others. Key steps are sampling, extraction, and detection.

### 6.2 Sampling

The choice of sampling is based on the properties of the analyte, environmental conditions, and the scope and goals of the study. When extracting samples from ambient air, the vapor pressure is the primary physical property that dictates its distribution as an aerosol or vapor, which are collected differently. Siting will depend on the topography and the meteorological conditions. Number of receptors and positioning are important considerations for being representative of the region in question. It is important to collect sufficient volumes of air to exceed the limit of quantitation of the analytical methods employed. Such characteristics will help in designing a sampling approach that is quantitative and representative.

## 6.2.1 Sampler Design

In 1960, the U.S. Public Health Service (USPHS) embarked on companion studies of pesticides in air and in human adipose tissues in the United States, mainly organochlorine (OC) insecticides and chlorinated contaminants like polychlorinated biphenyls (PCBs). The air study was contracted to the University of Miami, which developed and applied Greenburg–Smith impingers to trap and retain mainly OC and organophosphate (OP) pesticides (Figure 6.1) (Seiber et al., 1975). Principally OC pesticides were found in ambient air. However, GS impingers were fragile glass devices that used ethylene glycol as the trapping agent, and it was difficult to isolate and analyze pesticides in the presence of ethylene glycol (Figure 6.1). The air flow in impingers (max. 20 L/min) was too low for high-volume sampling, and the limit of detection (LOD) in impinger-based samplers was too high for analyzing trace levels of pesticides in air.

The ethylene glycol-filled impinger was replaced by an adsorption tube packed with a polystyrene/divinylbenzene macroreticular resin, Chromosorb 102, capable of trapping and retaining a variety of OC and OP pesticides and herbicides as well as nonpesticide contaminants (Thomas and Seiber, 1974). Woodrow et al. designed a battery-powered portable air sampler that operated at 20 L/min for up to 20 hours (Woodrow and Seiber, 1978). This team had effectively moved past the glycol impingers to a more efficient

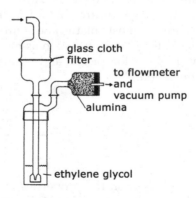

**FIGURE 6.1**
The Greenburg–Smith impinger was used in early studies to sample air for pesticides at EPA in Research Triangle Park and Perrine, Florida. (Reprinted/adapted by permission from Plenum Press: Springer: Seiber, J.N., Woodrow, J.E., Shafik, T.M., and Enos, H.F. (1975). Determination of pesticides and their transformation products in air. In *Environmental Dynamics of Pesticides*, (pp. 17–43) Springer. Copyright Plenum Press 1975.)

and rugged resin sampling tube which with modification is the backbone of modern air samplers for measuring ambient levels of pesticides in air. It can be adapted to smaller scale versions for assessing human exposures in a variety of conditions. When preceded by a glass fiber filter, one sampling tube could yield results for particulates and vapors. This device was used to assess exposures to pesticides in several use scenarios in fulfilling requirements of California's unique Toxic Air Contaminant Act (Baker et al., 1996). It was used to gather data on pesticide exposures needed for IR-4 registrations for minor use pesticides on specialty crops, as well as several other applications (Hengel and Lee, 2014).

But one size did not fit all pesticides and related contaminants of potential interest. Some highly volatile fumigants, such as methyl bromide and sulfuryl fluoride, were not trapped by XAD/Chromosorb resins requiring use of an alternative adsorbent—activated charcoal. Filters are typically made of glass or quartz fiber, and adsorbents are typically XAD, Chromosorb 100 series, Carbopack, Tenax, or silica gel, among others (Kosikowska and Biziuk, 2010). With these substitutions practically all pesticides could be analyzed by accumulative sampling in air.

## 6.2.2 Physical Properties of the Analyte

Methods for sampling of semivolatile organic compounds (SVOCs) including pesticides and their solvent carriers depend on the vapor pressure of the analyte (Woodrow et al., 2003) (Table 6.1). At vapor pressures (P) greater than 0.1 Pa, the pesticide is primarily in the vapor phase. Fumigants including methyl bromide (P = 216,645 Pa) and 1,3-D (P=3,866 Pa) and the herbicide S-ethyl dipropylthiocarbamate (EPTC) (P=4.5 Pa) are collected as vapors on adsorbents, adsorbent coatings, and impingers, or can be sampled from whole air in canisters. If the vapor pressure is less than $10^{-5}$ Pa, such as occurs for paraquat and the salts of many herbicides, they will be found primarily in the particulate or aerosol form. Aerosols are captured on filters and inertial

**TABLE 6.1**

Sampler Design Based on Vapor Pressure

|  | Example | Saturation Vapor Pressure (Pa) | Sampler Design |
|---|---|---|---|
| Vapors | Methyl bromide, S-ethyl dipropylthiocarbamate (EPTC) | $>10^{-1}$ | Adsorbents, canisters, impingers |
| Vapors+aerosols | Chlorinated hydrocarbons and organophosphates | $10^{-5}$ to $10^{-1}$ | Filters, adsorbents |
| Aerosols | Phenoxy herbicide salts and paraquat | $<10^{-5}$ | Filters, impactors, cyclone separators |

*Source:* Adapted from Woodrow et al. (2003).

samplers, such as impactors (cascade, dichotomous; see Chapter 7 on fog) and cyclone separators. If the vapor pressure is in the range of $10^{-5}$ to 0.1 Pa, both vapor and aerosol forms will be found in air, and a combination of the two sampling methods will be needed.

In most cases, samplers are equipped to capture both vapor and aerosols to assess an exposure event. For one, many pesticides are found in both vapor and aerosol phases. Also, the partitioning of chemicals between vapor and particulate phases depends on the particulate matter in the air. A high-volume air sampler, operating at $1\,m^3/min$, is depicted in Figure 6.2. The inlet is equipped with a filter to trap aerosols and particles followed by an adsorbent to trap vapor-phase chemicals. To avoid loss of analyte, e.g., through revolatilization from the filter or particulates in the filter, a vapor trap is positioned after the filter. All sampler components must be checked for sampling efficiency under conditions of use. This is commonly done by trapping experiments in the lab before field use, and also in field tests using spike recovery experiments. Sampling methods require calibration for both trapping and desorption efficiencies (Kosikowska and Biziuk, 2010).

Passive air samples are an alternative, particularly for high-volume cumulative air sampling. They do not require a pump and use media with high retention capacity (Kosikowska and Biziuk, 2010; Yusà et al., 2009). Such a passive sampler was used over the length of a year to monitor ambient air near a school in Kauai, Hawaii after incidences of exposure were reported (Wang et al., 2017). But passive air sampling does not usually yield quantitative data on concentrations of contaminants in air; at best they only provide estimates within a factor of 2–3 of the actual concentration (Yusa et al., 2009).

### 6.2.3 Cumulative Sampling

With cumulative sampling, vapors are trapped on a resin (e.g., XAD-4 or charcoal) or bead (e.g., glass) coated with a reactive chemical. Later, the analyte is extracted with solvents, or desorbed with heat. The fumigant, methyl isothiocyanate, is rapidly transformed to the more toxic methyl isocyanate (MIC) in the lower atmosphere. MIC was trapped on XAD-7 cartridges, coated with

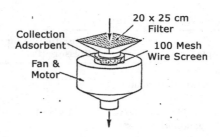

**FIGURE 6.2**
High-volume air sampler for particles and organic vapors in air. (Adapted from Hsieh et al., 1981.)

1-(2-pyridyl)-piperazine, which retains MIC as its stable substituted urea derivative. Formaldehydes were trapped on a polystyrene support coated with 2-(hydroxymethyl)-piperidine to form a stable oxazolidine derivative with formaldehyde (Hendricks, 1989). Another cumulative sampling method was developed by NIOSH: Method 2016 describes trapping formaldehyde on silica gel coated with 2,4-dinitrophenylhydrazine converting it to the hydrazone (NIOSH, 2003).

### 6.2.4 Special Sampler Designs

Cascade impactors are useful for fractionating suspended particles from air to specifically analyze lung-penetrating particles. Sampling for coronavirus and other biologic material can be done using the same type of filter used in protective facemasks or equivalent materials from 3M.

Analyzing fumigants and other volatiles in air when the analyte is in relatively high concentration can be as simple as using a gas tight syringe to withdraw a specified amount that can then be injected into a gas chromatograph (GC) or other analytical instrument. Whole air sampling can also be done with a Tedlar bag or deactivated canister.

Analysis of fumigated grains and spices can be done by headspace GC, a technique used for analyzing volatiles in wine and other beverages (Woodrow et al., 1995). Headspace GC and purge and trap are frequently used for analyzing volatile contaminants like solvents and fuels in water (Woodrow and Seiber, 1991). Pesticide vapors and other volatile chemicals can be sampled from the air, water, or other matrices with solid-phase extraction (SPE) or solid-phase microextraction (SPMEs) and then analyzed by gas or liquid chromatography, with or without mass spectrometry (MS) (Xiao et al., 2014).

### 6.2.5 Environment

Residues in air are generally highly mobile and relatively stable. Long-range transport occurs with many pesticides and contaminants such as PCBs and toxic air contaminants; they migrate over the ocean, to polar regions, or to mountain ranges long distances from the source. Such residues are often found in remote regions and in deposition associated with dust, rain, snow, and fog (see Chapters 7 and 10) (Kurtz, 1990). Monitoring fragile ecosystems, such as those in the Sierra Nevada Mountains, have surfaced numerous current-use pesticides, as summarized by the Sierra Nevada–Southern Cascades interagency study (Simonich and Nanus, 2012). Similar studies in the Mississippi River Valley found methyl parathion, parathion, diazinon, chlorpyrifos, endosulfan, malathion, etc. Pesticides were monitored along the Mississippi River in urban and agricultural sites, plus a background site near Lake Superior in Michigan. The herbicide atrazine, its transformation product, CIAT (2-chloro-4-isopropylamino-6-amino-s-triazine), and dacthal

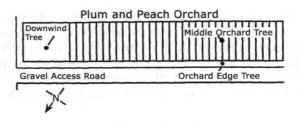

**FIGURE 6.3**
Potted pine seedlings placed in and around the orchard on the afternoon of spraying: in the
middle of the orchard, at the edge of the orchard, and 250 miles east of the orchard. (Reprinted
with permission from Aston, L. and Seiber, J. (1996). Exchange of airborne organophosphorus
pesticides with pine needles. *J. Environ. Sci. Health Part B* 31, 671–698. Copyright 1996 by Marcel
Dekker, Inc.)

were detected most frequently (76%, 53%, and 53%, respectively) at the background site indicating their propensity for long-range atmospheric transport
(Majewski et al., 2000).

Most air sampling is done at a one-meter elevation, but higher elevations
can be attained using samplers mounted on towers (Seiber et al., 1996). Pine
needles were used as passive samplers to assess pesticides in the air (Aston
and Seiber, 1996). Potted pine seedlings were arranged, as shown in Figure 6.3,
then pine needles were collected each afternoon for 7 days and then every
other day for the next 2 weeks. These "sentinel plants," once extracted and
analyzed, gave a good assessment of the pesticides/contaminants present in
a given airshed (also see Chapter 7).

## 6.3 Analysis

Once the samples are collected, contaminants can be extracted from the filter or trap separately and determined by MS-based techniques (Hengel and
Lee, 2014). In the past (before the advent of MS), extraction was followed by
fractionation, which usually involved normal phase (silica) chromatography,
collection of fractions, and analysis by GC of each fraction. MS can circum-
vent the need for fractionation. Nowadays, detection of analytes is best done
by capillary GC/MS or LC/MS in selective ion mode.

Among the techniques of extracting contaminants from trapping media are
solvent extraction, thermal extraction, accelerated solvent extraction, ultra-
sound assisted extraction, among others (Kosikowska and Biziuk, 2010). Newer
methods are typically aimed at reducing the amount of solvents used and
increasing throughput. Examples of these new methods include microwave-
(Coscollà et al., 2009) and sonication-assisted extraction (Nascimento et al.,
2018).

Volatile pesticides can be separated by capillary gas chromatography (GC) and detected by MS and tandem mass spectrometry (MS/MS) or the various other GC detectors including electron-capture detector, flame ionization detector, etc. Polar compounds can be detected with liquid chromatography and MS, MS/MS, or a UV detector. Alternatives include Fourier Transform spectroscopy (though it generally has a high LOD) and immunoassay (for targeted analysis).

Toxaphene is comprised of 175-plus discrete polychlorinated terpenes—primarily camphenes and bornanes, CxHyClz (PubChem, 2005). Toxaphene provides an example of sampling and analysis in air. The initial residue after application shows a GC profile characteristic of technical toxaphene (Figure 6.4a) after the residue weathers for 2 or more days; there is a clear shift in the fingerprint in favor of more volatile, less chlorinated congeners

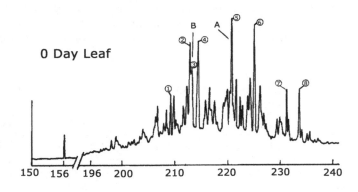

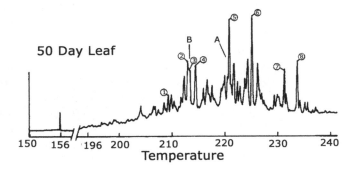

**FIGURE 6.4**

Capillary gas chromatograms of toxaphene residues on cotton leaves on day 0 and day 50. (Reprinted (adapted) with permission from Seiber, J.N., Madden, S.C., McChesney, M.M., and Winterlin, W.L. (1979). Toxaphene dissipation from treated cotton field environments: Component residual behavior on leaves and in air, soil, and sediments determined by capillary gas chromatography. *J. Agric. Food Chem.* 27, 284–291. Copyright 1979 American Chemical Society.)

because the more volatile components evaporate first from the residue on leaves or foliage. Volatilization represents the primary pathway for dissipation. By day 50, the GC profile shows a shift in favor of more chlorinated congeners that are less volatile (Figure 6.4b). The capillary GC profile shows no indication of chemical breakdown because no new peaks are seen in the weathered material.

## 6.4 Limit of Detection

The potential pitfalls associated with efficiency of sampling and post-sampling recovery must be taken into account in validating a sampling protocol (Woodrow et al., 2018). To ensure validity of the analytical methods used to measure pesticides in air, quality assurance and quality control procedures are required. It is important to consider the limits of detection and limits of quantitation for any sampling and analysis procedure (Table 6.2).

TABLE 6.2

Limit of Detection (LOD) and Limit of Quantitation (LOQ) for Pesticides Sampled from Air Using XAD-4 Adsorbent

|  | Average | SD[b] | LOD[c] | LOQ[d] |
| --- | --- | --- | --- | --- |
| Compound[a] | (µg/sample) | (µg/sample) | (µg/sample) | (µg/sample) |
| Chlorothalonil | 0.089 | 0.011 | 0.032 | 0.161 |
| Chlorpyrifos | 0.098 | 0.006 | 0.017 | 0.087 |
| Chlorpyrifos oxon | 0.101 | 0.004 | 0.012 | 0.061 |
| Diazinon | 0.093 | 0.005 | 0.016 | 0.081 |
| Diazinon oxon | 0.097 | 0.004 | 0.012 | 0.059 |
| Dimethoate | 0.097 | 0.004 | 0.012 | 0.062 |
| Dimethoate oxon | 0.102 | 0.004 | 0.011 | 0.053 |
| EPTC | 0.092 | 0.004 | 0.014 | 0.069 |
| Fonofos | 0.09 | 0.005 | 0.015 | 0.074 |
| Fonofos oxon | 0.094 | 0.004 | 0.012 | 0.06 |
| Malathion | 0.098 | 0.006 | 0.019 | 0.093 |
| Malathion oxon | 0.102 | 0.003 | 0.009 | 0.045 |
| Metolachlor | 0.111 | 0.004 | 0.013 | 0.065 |
| Permethrin | 0.112 | 0.011 | 0.032 | 0.161 |
| Simazine | 0.109 | 0.004 | 0.014 | 0.068 |
| Trifluralin | 0.114 | 0.012 | 0.034 | 0.171 |

*Source:* Adapted from Hengel and Lee (2014).
[a] Determined during 2,000 sampling projects at 0.1 µg/sample.
[b] Standard deviation ($n=8$).
[c] LOD is t value (2.998 for $n=8$) × standard deviation.
[d] LOQ is LOD×5.

Pesticides are typically in "trace" amounts in the air. Inadequately designed studies may result in "not detected" or "less than" values. For viruses and other biologics, a surrogate material of similar size and other characteristics may be needed to evaluate trapping and extraction efficiency—lentovirus has been suggested as a nontoxic surrogate for COVID-19.

## 6.5 Further Reading

More detailed information on analyzing pesticides and related toxicants in air is in the excellent review by Woodrow et al. (2003). An older review is by Lewis and Lee (1976).

## References

Aston, L., and Seiber, J. (1996). Exchange of airborne organophosphorus pesticides with pine needles. *J. Environ. Sci. Health Part B* 31, 671–698.

Baker, L.W., Fitzell, D.L., Seiber, J.N., Parker, T.R., Shibamoto, T., Poore, M.W., Longley, K.E., Tomlin, R.P., Propper, R., and Duncan, D.W. (1996). Ambient air concentrations of pesticides in California. *Environ. Sci. Technol.* 30, 1365–1368.

Coscollà, C., Yusà, V., Beser, M.I., and Pastor, A. (2009). Multi-residue analysis of 30 currently used pesticides in fine airborne particulate matter (PM 2.5) by microwave-assisted extraction and liquid chromatography–tandem mass spectrometry. *J. Chromatogr. A* 1216, 8817–8827.

Hendricks, W. (1989). Acrolein and/or Formaldehyde Method 52. Occupational Safety & Health Administration. https://www.osha.gov/dts/sltc/methods/organic/org052/org052.html (Accessed June 4, 2021).

Hengel, M., and Lee, P. (2014). Community air monitoring for pesticides—part 2: multiresidue determination of pesticides in air by gas chromatography, gas chromatography–mass spectrometry, and liquid chromatography–mass spectrometry. *Environ. Monit. Assess.* 186, 1343–1353.

Hsieh, D.P.H., Seiber, J.N., and Fisher, G.L. (1981). *Final Report Agreement Number: AB-O93-31 California Air Resources Board May 1981*. Davis, CA: California Air Resources Board.

Kosikowska, M., and Biziuk, M. (2010). Review of the determination of pesticide residues in ambient air. *TrAC Trends Anal. Chem.* 29, 1064–1072.

Kurtz, D.A. (1990). *Long Range Transport of Pesticides*. Chelsea, MI: Lewis Publishers, Inc.

Lewis, R.G., and Lee, R.E. (1976). *Air Pollution from Pesticide and Agricultural Processes*. Cleveland, OH: CRC Press.

Majewski, M.S., Glotfelty, D.E., and U, K.T.P. (1990). A field comparison of several methods for measuring pesticide evaporation rates from soil. *Environ. Sci. Technol.* 24, 1490–1497.

Majewski, M.S., Foreman, W.T., and Goolsby, D.A. (2000). Pesticides in the atmosphere of the Mississippi River Valley, part I — rain. *Sci. Total Environ.* 248, 201–212.

Nascimento, M.M., da Rocha, G.O., and de Andrade, J.B. (2018). A rapid low-consuming solvent extraction procedure for simultaneous determination of 34 multiclass pesticides associated to respirable atmospheric particulate matter (PM2. 5) by GC–MS. *Microchem. J.* 139, 424–436.

NIOSH. (2003). Method 2016: Formaldehyde. Issue 2. *NIOSH Man. Anal. Methods Third Suppl.* DHHS (NIOSH) Publication No. 2003-154. Washington, DC.

National Center for Biotechnology Information. (2021). PubChem Compound Summary for CID 5284469, *Toxaphene*. Retrieved June 4, 2021 from https://pubchem.ncbi.nlm.nih.gov/compound/Toxaphene.

Seiber, J.N., Madden, S.C., McChesney, M.M., and Winterlin, W.L. (1979). Toxaphene dissipation from treated cotton field environments: Component residual behavior on leaves and in air, soil, and sediments determined by capillary gas chromatography. *J. Agric. Food Chem.* 27, 284–291.

Seiber, J.N., Woodrow, J.E., Shafik, T.M., and Enos, H.F. (1975). Determination of pesticides and their transformation products in air. In *Environmental Dynamics of Pesticides*, (pp. 17–43). Boston, MA: Springer. Haque R., Freed V.H. (eds).

Seiber, J.N., Woodrow, J.E., Yates, M.V., Knuteson, J.A., Wolfe, N.L., and Yates, S.R. (1996). *Fumigants: Environmental Fate, Exposure, and Analysis*. Washington, D.C.: ACS Publications.

Simonich, S.L., and Nanus, L. (2012). Sierra Nevada-Southern Cascades (SNSC) region air contaminants research and monitoring report. Sierra Nev.-South. Cascades Steering Committee.

Thomas, T.C., and Seiber, J.N. (1974). Chromosorb 102, an efficient medium for trapping pesticides from air. *Bull. Environ. Contam. Toxicol.* 12, 17–25.

Wang, J., Boesch, R., and Li, Q.X. (2017). A case study of air quality - Pesticides and odorous phytochemicals on Kauai, Hawaii, USA. *Chemosphere* 189, 143–152.

Woodrow, J.E., Gibson, K.A., and Seiber, J.N. (2018). Pesticides and Related Toxicants in the Atmosphere. In P. de Voogt (Ed.) *Reviews of Environmental Contamination and Toxicology* (Vol. 247, pp. 147–196.) Cham: Springer International Publishing.

Woodrow, J.E., Hebert, V., and LeNoir, J.S. (2003). Monitoring of agrochemical residues in air. *Handb. Residue Anal. Methods Agrochem. Ed Philip W Lee* 1, 908–935.

Woodrow, J.E., McChesney, M.M., and Seiber, J.N. (1995). Determination of ethylene oxide in spices using headspace gas chromatography. *J. Agric. Food Chem.* 43, 2126–2129.

Woodrow, J.E., and Seiber, J.N. (1978). Portable device with XAD-4 resin trap for sampling airborne residues of some organophosphorus pesticides. *Anal. Chem.* 50, 1229–1231.

Woodrow, J.E., and Seiber, J.N. (1991). Two chamber methods for the determination of pesticide flux from contaminated soil and water. *Chemosphere* 23, 291–304.

Xiao, L., Lee, J., Zhang, G., Ebeler, S.E., Wickramasinghe, N., Seiber, J., and Mitchell, A.E. (2014). HS-SPME GC/MS characterization of volatiles in raw and dry-roasted almonds (Prunus dulcis). *Food Chem.* 151, 31–39.

Yusà, V., Coscollà, C., Mellouki, W., Pastor, A., and De La Guardia, M. (2009). Sampling and analysis of pesticides in ambient air. *J. Chromatogr. A* 1216, 2972–2983.

# 7

## Pesticides in Fog

### 7.1 Introduction

Fog has long played a historical and mythical role in human history, in fog enshrouded coast lines that inspired creation of light houses to guide mariners, to sophisticated radar systems on board aircraft and ocean fleets to avoid collisions. There are numerous fog-related accidents that annually claim lives in highway, train, and shipping lanes. In WWII, fog played a vital role in the Battle of Britain, illustrated well in allied bombing raids over Germany near the end of Nazi Germany's domination in the European theater—their sustained success relied on the safe return of allied bombers over the cliffs of Dover to landing strips often fogged into the west of Dover. A network of shallow canals or berms containing combustible aircraft fuel was used to burn off the fog layer and outline landing strips for pilots returning in battered airplanes running near empty of fuel.

In this chapter, we describe research on pesticides and related contaminants in fog prone food production areas of California's San Joaquin and other coastal valleys (Figure 7.1). The partitioning of chemicals from air into water can extend to all forms of water, including fogwater. Therefore, it is no surprise fog and rainwater can accumulate residues, and thus help cleanse the atmosphere of contaminants.

### 7.2 Pesticides in Fogwater

In the winter of 1970, Dwight Glotfelty of USDA-Agricultural Research Station (ARS) and James Seiber of UC, Davis were sampling and analyzing water bodies (rainwater, streams, irrigation water, lakes, etc.) for pesticide residues near Chesapeake Bay near Beltsville, Maryland, and in the San Joaquin Valley. Fog was not common in the Chesapeake Bay area, and some early, rudimentary attempts to sample fogwater there largely failed. However, they hypothesized that areas with heavy fog, such as the tule fog

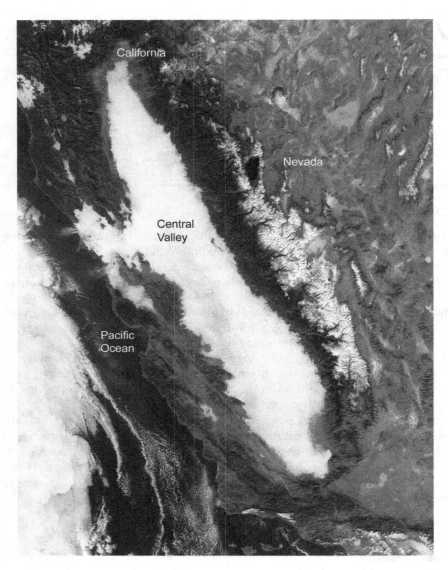

**FIGURE 7.1**
Aerial photo of tule fog blanket in the California Central Valley, prevalent during winter months (December–February). The high humidity, calm winds, and rapid nighttime cooling in the valley develops the tule fog. The long winter nights cool the ground and produce a temperature inversion at low altitudes. (Credit to Jeff Schmaltz, National Aeronautics and Space Administration (NASA).)

that occurs in California's San Joaquin Valley, could capture high concentrations of pesticides—if they could only figure out a way to sample it. Enter the Ag Engineers at USDA Beltsville (Lou Liljedahl and others), who came up with a novel sampling mechanism mounted on the cab of a pickup truck. With an additional assist from the U.S. Air Force, who flew the truck from Maryland to Fairfield, California, the researchers now had a way to collect sizeable volumes of fog easily. All they had to do was drive around the orchards and agricultural fields during fog events.

Their unusual work quickly grabbed attention. Using gas chromatography (GC) analytical techniques with element selective detectors, Seiber, Glotfelty, Woodrow, McChesney, Lucas, and other researchers at UC, Davis, and USDA ARS identified peaks for pesticides, polycyclic aromatic hydrocarbons (PAHs), phthalate esters, organophosphate (OP) esters, herbicides, and organochlorine (OC) insecticides. The local media highlighted their work and one headline, titled "Killer Fogs," fueled interest in the project. Additional media coverage and publications fueled a multiyear study to understand contaminants in fog.

Wet deposition, which includes the scavenging of particle-bound pesticides and pesticide vapors into atmospheric moisture (cloud- and fogwater, rain, and snow), is a potentially major sink for airborne pesticides. The pervasive wintertime tule fogs in California's Central Valley, studied extensively in the past 25 years, accumulate organophosphorus, triazine, and other pesticide groups. Concentrations of some pesticides in fogwater can significantly exceed those expected based upon vapor–water distribution coefficients. Fogwater deposition has been implicated as a source of inadvertent residues to nontarget foliage, and of high-risk exposures for wildlife residing in and around treated areas. The pesticide residue content of fogwater is an example.

Airborne pesticides may exist as vapors or associated with liquid or solid aerosols (Lewis and Lee, 1976). Vapor–aerosol distribution and partitioning of airborne pesticides will occur, as it does with all volatile and semivolatile air contaminants (Bidleman, 1988). Liquid-phase vapor pressure is the controlling physical property. In general, chemicals of vapor pressures less than about $10^{-10}$ atmospheres favor the particulate phase and those with vapor pressures greater than about $10^{-9}$ atmospheres favor the vapor phase. Pesticides cover a broad range of vapor pressure, extending from gaseous chemicals under ambient conditions (e.g., methyl bromide, sulfuryl fluoride, and phosphine) to essentially nonvolatile salts such as paraquat. But the majority of pesticides are semivolatile organic compounds (SVOCs) with vapor pressure falling in the range of, approximately, $10^{-4}$ to $10^{-10}$ atmospheres.

Airborne pesticides may be removed from the air by principally three processes, namely, degradation, wet deposition, and dry deposition. Degradation may follow oxidative, photooxidative, and/or hydrolytic pathways. Only a few pesticide chemicals have been studied in detail, and they display the same types of reactions seen for other classes of organics in the air (see Chapter 3) (Seiber and Woodrow, 1995; Woodrow et al., 1983). In only a very few cases are atmospheric vapor transformations rapid enough to be of significance in the time scale of minutes or hours; such cases include, merphos oxidation to S,S,S-tributyl phosphorotrithioate (DEF), OP thion oxidation to the corresponding oxon, and trifluralin N-dealkylation. At the other end of the spectrum are chemicals such as dichlorodiphenyldichloroethylene (DDE) and methyl bromide. Methyl bromide has a tropospheric lifetime of approximately 1 year, which allows this chemical to diffuse, unreacted, to the stratosphere (Yvon-Lewis and Butler, 1997).

Dry deposition involves the settling of particles, which is strongly influenced by particle size and the nature of the meteorology and terrain, and by direct vapor-surface exchange. Wet deposition includes the scavenging of particle-bound pesticides and pesticide vapors into atmospheric moisture (cloud- and fogwater, rain, and snow), followed by rainfall, snowfall, or fog droplets coalescing on surfaces. This is potentially a major sink for airborne pesticides, a source of exposure to pesticides for vegetation, aquatic organisms, and watershed ecosystems, and a means of degrading hydrolytically labile airborne pesticides. The content of pesticides in rain, cloud, fog, and snow has been studied extensively only in the past 25 years (Rice, 1996). The accumulating information is quite compelling. Pesticides are measurably present in air and rainfall sampled throughout the United States (Goolsby et al., 1994; Majewski and Capel, 1995). Pesticides are also found in snow and ice, including in remote regions of the earth (Kurtz, 1990). And pesticides used (emitted) in the southeast or southern United States ride the storm fronts presenting major deposition inputs to the Chesapeake Bay (Glotfelty et al., 1990a), Great Lakes (Eisenreich et al., 1981), and other water bodies. Pesticides and other anthropogenic trace organics are found in cloud- and fogwater where they may achieve concentrations even greater than those expected based upon vapor–water distribution calculations (Glotfelty et al., 1987).

Fogwater residues in particular have been implicated as sources of inadvertent residues to nontarget crops (Turner, 1989) and of high-risk exposures for hawks residing around treated orchards and for pumas and other predators which frequent water sources in coastal areas which receive marine salt water sprays from the surf (Wilson et al., 1991). Fogwater is also an indicator of long-range transport of pesticides to remote regions of the earth (Chernyak et al., 1996). The present chapter will delve into the data and underlying methodology associated with these findings.

## 7.3 Pesticide Use in California

The coastal valleys of California, including the Central Valley, receive extensive year-round use of a variety of pesticides and frequent occurrence of wintertime fogs. The Central Valley comprises a land mass of nearly 50,000 km² residing between California's coastal mountain range on the west, the Sierra Nevada Mountain range foothills on the east, and extending roughly from the cities of Redding (north) to south of Bakersfield. The northern portion, the Sacramento Valley, is dominated by rice and wheat as major field crops, and almond and other fruit and nut orchards. The southern portion, the San Joaquin Valley, produces cotton, wheat, corn, and a variety of vegetables (carrots, broccoli, cabbage, melons) as major field crops while grape, citrus, almond, walnut, nectarine, peach, plum, and apricot are among the major vineyard and orchard crops. Pesticide use in California is extensive (see Table 2.1 for the leading pesticides used in California in 2018). Soil fumigants (methyl bromide, metam sodium, chloropicrin), OP and chlorinated hydrocarbon insecticides-fungicides, triazine and thiocarbamate herbicides, cotton defoliants such as sodium chlorate, paraquat, and DEF (S,S,S-tributyl phosphorotrithioate), and many others are included. Fog episodes in the Central Valley occur predominately in the November–February period. This period coincides with the dormant spraying of fruit and nut orchards. These sprays include typically sulfur, mineral oil, OP insecticides, and some other chemicals (Steinheimer et al., 2000).

## 7.4 Fogwater Sampling Methodology

Seiber and Glotfelty began their fog research using the Caltech Rotating Arm Collector (RAC) (Figure 7.2), which featured a 1.5-HP motor to drive a 63 cm long stainless steel rod at 1,700 rpm through the air (Jacob et al., 1984). Each end of the rod had a slot milled into its leading edge to collect impacting fogwater droplets. High-density polyethylene bottles mounted on each end of the rod catch the fogwater samples as they are accelerated from each slot. Collection rates up to 2 mL/min were obtained in the field (Collett et al., 1990). They soon transitioned to a new design—a high-volume collector developed at USDA-ARS laboratory in Beltsville, MD (Glotfelty et al., 1986). The innovation was a rotating screen mechanism, 50 cm in diameter, in which four layers of stainless steel screen material were rotated around a central axis at 720 rpm. Fogwater obtained from droplets impacting on the screen was centrifuged to the periphery, collected in a slotted aluminum tube, and drained into a collection vessel. A large fan pulled air through the device at a sampling rate of ~160 m³/min. The sampling rate typically allowed collections of approximately 1 L/hr of fogwater depending on the suspended water content of the fog. This device was employed mounted on a pickup truck, which moved slowly (5–15 km/hr) through the foggy area to be sampled. Vehicle and gas-powered generator fumes were thus continually swept toward the rear and away from the fog sampler intake.

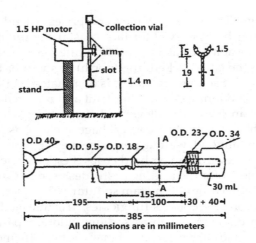

**FIGURE 7.2**
Rotating arm collector. (Reprinted from Collett Jr, J.L., Daube Jr, B.C., Munger, J.W., and Hoffmann, M.R. (1990). A comparison of two cloudwater/fogwater collectors: The rotating arm collector and the caltech active strand cloudwater collector. *Atmospheric Environ. Part Gen. Top.* 24, 1685–1692. Copyright (1990), with permission from Elsevier.)

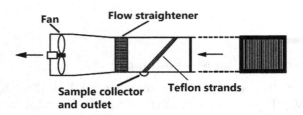

**FIGURE 7.3**
CalTech Active Strand cloudwater collector—currently the most popular design, in use by DPR. (Reprinted from Collett Jr, J.L., Daube Jr, B.C., Munger, J.W., and Hoffmann, M.R. (1990). A comparison of two cloudwater/fogwater collectors: The rotating arm collector and the caltech active strand cloudwater collector. *Atmospheric Environ. Part Gen. Top.* 24, 1685–1692. Copyright (1990), with permission from Elsevier.)

Today, a common fog collector is the Caltech Active Strand Cloudwater Collector (CASCC) (Figure 7.3), which employs a fan to draw air across six angled teflon banks at a velocity of 8.45 m/sec (Daube, 1987). Cloud or fogwater droplets in the air are collected by the strands by inertial impaction. The collected droplets run down the strands, aided by gravity and aerodynamic drag, through a Teflon sample trough into a collection bottle. This instrument has a theoretical lower size-cut of 3.5 μm, based on droplet diameter, and has collected fog- and cloudwater at rates of up to 8.5 mL/min in the field (Collett et al., 1990).

A useful adjunct to any fog- and cloudwater collection is the availability of a method for determining the presence and density of these atmospheric water suspensions. Mallant and Kos described a low-cost optical fog detector which

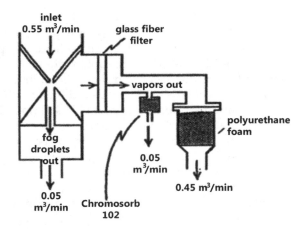

**FIGURE 7.4**
High-volume dichotomous impactor for sampling the interstitial pesticide vapors in fog.
(Adapted from Glotfelty et al. 1986.)

had many advantages over other detection devices described in the literature
(Mallant and Kos, 1990). Fog sampling for pesticide analysis in the Central
Valley was generally accompanied by simultaneous collection of interstitial
air sampled by means of a high-volume dichotomous sampler (Figure 7.4)
(Glotfelty et al., 1986). This provided a sample of the interstitial vapor- and
particle-phase pesticides by eliminating fog droplets of >8 μm through the
large particle orifice of this device. Fog droplets of <8 μm, unactivated (dry)
aerosol particles and vapor passed first through a glass fiber filter (GFF), which
removed particulate matter, and then through a 7.5 cm diameter × 7.5 cm deep
bed of porous polyurethane foam (PUF) which trapped the pesticide vapors.
The PUF plugs were precleaned by the method of Bidleman and Olney (1975).
Alternatively, a Chromosorb 102 trap was used to collect vapor samples at
~1 L/min, for vapor analysis (Thomas and Seiber, 1974). Extracts of fogwater, particle filters, and vapor traps were analyzed by GC directly, or following fractionation on a silica High Pressure Liquid Chromatography (HPLC)
column using a hexane-to-methyl t-butylether (MTBE) gradient (Seiber et al.,
1990; Wehner et al., 1984). Major components were identified by GC retention
time, HPLC retention behavior, and GC–MS, in comparison with authentic
samples. XAD or Chromosorb 102 were used to trap vapors for pesticides.

## 7.5 Fogwater Sampling Results

Using the USDA rotating screen fogwater collector and dichotomous air sampler, Glotfelty and Seiber et al. presented the landmark studies of pesticides

in fog (Glotfelty et al., 1986, 1987). They discovered that a variety of pesticides and their toxic alteration products are present in fog, and that they can reach high concentrations relative to rainwater and other atmospheric water samples. Also, in measuring the air–water distribution of pesticides between the suspended liquid phase in fogwater and the interstitial air, they found that some chemicals are enriched several thousand-fold in the suspended liquid when compared with distributions expected based upon Henry's law coefficients for pure air–water distribution systems.

These fog samples were collected at the USDA Beltsville Agricultural Research Center in Beltsville, MD, and at several locations in California's San Joaquin Valley. Beltsville samples were often dark in color, with large amounts of black particles (largely carbonaceous) which settled out when left to stand, leaving an off-yellow supernatant. Several common pesticides such as malathion were present in measurable levels. Industrial organophosphorus chemicals, polycyclic aromatic hydrocarbons, and phthalate esters were present in these samples as well. Although the USDA Beltsville Center is agricultural, it is surrounded by a densely populated, partially industrial area with heavy vehicular traffic.

Table 7.1 lists the pesticides and alteration products identified in fog in California (Glotfelty et al., 1987). Organophosphorus insecticides (diazinon, parathion, chlorpyrifos, methidathion, malathion, methyl parathion) and

**TABLE 7.1**

Distribution of Pesticides and Their Alteration Products between Fogwater and Interstitial Air in California

| Chemical | Location | Chemical Concentration | | Distribution Ratio ($\times 10^6$) Air–Water | | |
| --- | --- | --- | --- | --- | --- | --- |
| | | Fogwater (ng/L) | Air (ng/m³) | Measured (D) | Literature (H) | Enrichment Factor (EF) |
| *OP Insecticides* | | | | | | |
| Diazinon | P | 16,000 | 2.2 | 0.12 | 60 | 500 |
| | C | 11,800 | 4.2 | 0.36 | | 170 |
| | L | 22,000 | 5.3 | 0.24 | | 249 |
| Parathion | P | 12,400 | 3.2 | 0.25 | 9.5 | 38 |
| | C | 5,800 | 0.88 | 0.15 | | 63 |
| | L | 51,400 | 7.9 | 0.15 | | 63 |
| Chlorpyrifos | P | 1,020 | 3.3 | 3.2 | 500 | 156 |
| | L | 320 | 0.6 | 1.9 | | 260 |
| | C | 6,500 | 14.7 | 2.3 | | 217 |
| Methidathion | P | 840 | <0.03 | <0.04 | 0.07 | >1.8 |
| | L | 570 | <0.01 | <0.02 | | >3.5 |
| | C | 15,000 | <0.6 | <0.04 | | >1.8 |

*(Continued)*

**TABLE 7.1** (*Continued*)

Distribution of Pesticides and Their Alteration Products between Fogwater and Interstitial Air in California

| Chemical | Location | Chemical Concentration | | Distribution Ratio ($\times10^6$) Air–Water | | Enrichment Factor (EF) |
|---|---|---|---|---|---|---|
| | | Fogwater (ng/L) | Air (ng/m³) | Measured (D) | Literature (H) | |
| Malathion | P | 70 | <0.03 | <0.4 | 2.4 | >6 |
| | L | 110 | 0.02 | 0.18 | | 13 |
| | C | 350 | | | | |
| *Oxygen Analogs (OAs)* | | | | | | |
| Parathion OA | P | 9,000 | 0.21 | 0.02 | 0.25 | 11 |
| | C | 950 | 0.66 | 0.07 | | 3.6 |
| | L | 184,000 | 0.25 | 0.00 | | 250 |
| Methidathion OA | P | 120 | 0.03 | <0.25 | – | – |
| | L | 8,200 | | | | |
| Chlorpyrifos OA | P | 170 | <0.03 | <0.18 | – | – |
| | L | 800 | | | | |
| Diazinon OA | P | 190 | <0.03 | <0.16 | – | – |
| *Cotton Defoliant* | | | | | | |
| DEF | P | 250 | <0.03 | <0.7 | 0.2 | – |
| | C | 800 | 0.14 | <0.4 | | |
| *Herbicides* | | | | | | |
| Atrazine | P | 270 | <0.2 | <0.7 | 0.2 | – |
| | C | 320 | <0.16 | <0.4 | | – |
| | L | 700 | – | – | | – |
| Simazine | P | 390 | <0.1 | <0.3 | 0.025 | – |
| | C | 110 | <0.06 | <0.5 | | – |
| | L | 1,200 | – | – | | – |
| Pendimethalin | P | 1,370 | 0.64 | 0.47 | 1,500 | 3,200 |
| | C | 3,620 | 3.6 | 1.0 | | 1,500 |

*Source:* Glotfelty, D.E., Seiber, J.N., and Liljedahl, A. (1987). Pesticides in fog. *Nature* 325, 602–605.

*Note:* Samples were collected in 1985 and parathion was banned by EPA in about 1980.
C, Corcoran, CA on January 13, 1985; EF, ratio of literature to measured distribution ratio, reflects aqueous-phase enrichment; L, Lodi, CA on January 19, 1985, P, Parlier, CA on January 13, 1985.

their oxygen analogs (OAs) were most frequently found. But several types of herbicides, including the triazines (atrazine, simazine), dinitroaniline (pendimethalin), and chloroacetanilides (alachlor, metalochlor), were measurably present in some samples. The fog samples from California yielded more variety and higher concentrations of pesticides than the Beltsville samples, a result of the greater diversity and heavier volume of pesticide use in California's Central Valley. The OP insecticides frequently exceeded 10 µg/L in fog, which is 2–3 orders of magnitude greater than for these and similar compounds in rain from other locations, and in the Central Valley itself (Seiber et al., 1993). Because these California samples were collected during wintertime dormant spraying in the Central Valley, the elevated levels of diazinon, methidathion, and chlorpyrifos were ascribed to this usage.

The California samples contained a number of OAs of OP pesticides, with parathion OA (paraoxon) being the most abundant. The source of parathion and paraoxon compounds is unknown. OAs form in the gas phase, by reaction of parent thions with atmospheric oxidants (Woodrow et al., 1983), and also on surfaces (Spear et al., 1975; Woodrow et al., 1977). OAs are potent inhibitors of cholinesterase and are generally responsible for most of the toxic effects of OPs (Henderson et al., 1994). The fog collected in Lodi, which had the highest measured concentration of paraoxon (184 µg/L), yielded a lab-measured cholinesterase inhibition which matched that of equivalent concentrations of pure paraoxon standard.

The Henry's law constant, which generally describes the distribution of low solubility organics between the vapor- and aqueous phase in equilibrium (Suntio et al., 1988), was not a good quantitative indicator of the distribution of pesticides between the vapor- and liquid phase in these fog samples. The measured air–water distribution coefficient (D) was very much less than H (Table 7.1). D<H implies an aqueous-phase enrichment; more pesticide is dissolved in the aqueous phase than would be expected in an ideal solution at equilibrium. The extents of the aqueous phase enrichment were given as enrichment factors (EF=H/D), which were quite large, even several thousand-fold, for chemicals such as pendimethalin and diazinon. The possible underlying reasons for enrichment, and its consequences, are discussed below.

The distribution of pesticides in foggy Central Valley atmospheres was studied in more depth (Glotfelty et al., 1990b). This study was conducted at one Central Valley sampling site (Kearney Agricultural Research Center, University of California, Parlier, CA) and included one foothill site about 50 km east of Parlier, at approximately 500 m elevation. The distribution of four OP insecticides (diazinon, parathion, chlorpyrifos, and methidathion) and their oxons between the droplet and air phases was studied during six fog events. Up to 50 µg/L was found for the total of the four OPs and up to 75 µg/L for the total of their oxons in the fogwater. Nearly all of the compounds exhibited aqueous phase enrichment, ranging from a mean of 1.4 (methidathion) to 58 (diazinon) (Table 7.2).

The oxon to thion ratios in the fogwater were generally less than 1, with the exception of two sampling sites and dates. Atmospheric oxidation, especially during daylight hours, followed by uptake in the fogwater was implicated. Even though there were high concentrations and high EF for the water phase, the very low volume of suspended water in fog leads to the highest

**TABLE 7.2**

Aqueous-Phase Enrichment Factors ($K_{AW}/D$) for the Distribution of
Pesticides between Fogwater and Air

| Compound | 1/8–9 | 1/9 | 1/11[a] | 1/12 | 1/12–13 | Mean |
|---|---|---|---|---|---|---|
| Diazinon | 6 | 59 | 14 | 50 | 160 | 58 |
| Parathion | 4 | 12 | 5 | 18 | 29 | 14 |
| Chlorpyrifos | 7 | 55 | 3.5 | 40 | 74 | 42 |
| Methidathion | 0.06 | 3 | 0.02 | 1.4 | 2.3 | 1.4 |
| Paraoxon | 2.1 | 14 | 10 | 48 | >69 | >19 |

*Source:* Reprinted (adapted) with permission from Glotfelty, D.E., Majewski,
M.S., and Seiber, J.N. (1990b). Distribution of several organophosphorus
insecticides and their OAs in a foggy atmosphere. *Environ. Sci. Technol.*
24, 353–357. Copyright (1990) American Chemical Society.

[a] Hills Valley Road site.

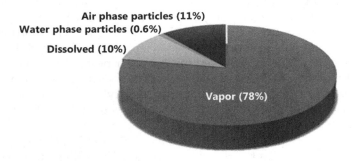

**FIGURE 7.5**
Distribution of parathion in the foggy atmosphere near Parlier, CA, on January 12, 1986.
(Reprinted (adapted) with permission from Glotfelty, D.E., Majewski, M.S., and Seiber, J.N.
(1990b). Distribution of several organophosphorus insecticides and their OAs in a foggy atmosphere. *Environ. Sci. Technol.* 24, 353–357. Copyright (1990) American Chemical Society.)

proportion of all of the compounds, in all of the events, in the interstitial air
phase, either as vapor or adsorbed to aerosol particles.

Figure 7.5 shows this distribution for parathion for a single sample. The
total concentration of parathion in the atmosphere was 9.4 ng/m³. Of this,
78% (7.3 ng/m³) was in the vapor phase. Only 10% (0.9 ng/m³) was dissolved
in the fog droplet, even though the concentration in the droplet was high—
30 µg/L—resulting from the high enrichment factor (Glotfelty et al., 1987)
and low concentration (<0.1 mg/m³) of suspended water droplets in foggy
atmospheres. Up to 11% (1.0 ng/m³) appeared to be attached to aerosol particles. And only 0.6% was attached to solids filterable from the fogwater. It
was not determined whether the particulate-bound pesticides were present
adsorbed to particles before the fog formed or were sorbed out from dissolved pesticides present in the droplets after the fog formed.

In order to determine whether the uptake of pesticides in fog was unique to the wintertime ground fog, or tule fog, of the inland Central Valley, Schomburg et al. (1991) used a scaled-up teflon strand fog collector (Seiber et al., 1993) to analyze air and fogwater in several spring advective oceanic fogs collected near Monterey, CA, and the heavily agricultural Salinas Valley and associated coastal plains. The pesticide content and distribution for several pesticides common to both areas (chlorpyrifos, diazinon, etc.) were remarkably similar between the coastal and Central Valley samples. The conversion of thion to oxon and the aqueous phase enrichments were also very similar. The data also provided support for the hypothesis that nonfilterable, strongly sorptive particles and colloids in the fogwater cause the enhancement of pesticides in water in the foggy atmosphere.

Fogwater entrainment and concentration of chemicals are not unique to pesticides. Sagebiel and Seiber (1993) analyzed wintertime fog and interstitial air from a community in the Central Valley during a time when residential wood burning occurred. Guaiacol, 4-methyl guaiacol, and syringol were the most commonly found among the 16 methoxylated phenolic lignin combustion products confirmed by GC/MS in fog samples. The distribution of methoxylated phenols generally followed Henry's law, that is, did not show the dramatic enrichments observed for less polar pesticides. This suggested that enrichment is a function of analyte hydrophobicity rather than any special structural features. Concentrations of methoxylated phenols in fogwater ranged to 1,408 µg/L for syringol, and generally were in the 1–100 µg/L range when detected.

Seiber et al. (1993) provided more details on the distribution of pesticides in air, fogwater, and plant surfaces in a series of experiments carried out at the Kearney Agricultural Research Center at Parlier, CA. Fogwater contained residues of the four OP dormant spray insecticides (parathion, chlorpyrifos, diazinon, and methidathion) and their oxons (Table 7.3).

Concentrations ranged to 91 µg/L for parathion and 76 for diazinon and were significantly lower for oxons (to 19 µg/L for paraoxon, 6.2 µg/L for methidathion oxon, 3.4 µg/L for chlorpyrifos oxon, and 3.0 µg/L for diazinon oxon), in fogwater sampled with the Teflon strand collector. When fogwater was collected by simply placing a collector beneath the drip lines of tree canopies, similar water concentrations were observed, but the ratio of oxon to thion increased dramatically for the four insecticides, to an average of 0.7 (diazinon) and 1.35 (methidathion) (Table 7.3). This indicated either that conversion of thion to oxon occurred in fogwater as it collected on, and passed through the foliage, or that conversion of thion to oxon occurred in the tree parts (leaves, needles, limbs) after the water evaporated as the fog lifted, with the surface-formed oxon being removed during the next episode of fog. The apparent tree surface catalysis of thion to oxon was noted previously (Spear et al., 1975).

The phenomenon of enrichment of chemical solutes in the suspended aqueous phase of foggy atmosphere has been the subject of much discussion since

**TABLE 7.3**

Oxon/Thion Ratios in Fogwater Collected Using Active Strand Fogwater Sampler, as Tree Drip Beneath Three Types of Trees, and Non-fog, Clear Afternoon Air (Seiber et al., 1993)

|  | Chlorpyrifos | Diazinon | Parathion | Methidathion |
|---|---|---|---|---|
| Clear afternoon air | 0.45 | 0.52 | 0.62 | 0.58 |
| Fogwater collected by active strand sampling | 0.21 | 0.07 | 0.60 | 1.8 |
| Pine tree drip water | 1.13 | 0.86 | 1.11 | 1.43 |
| Deciduous tree drip water | 1.08 | 0.64 | 0.71 | 1.15 |
| Evergreen tree drip water | 1.27 | 0.59 | 0.83 | 1.46 |
| Tree drip water average | 1.16 | 0.70 | 0.88 | 1.35 |

*Source:* Reprinted (adapted) with permission from Seiber, J.N., Wilson, B.W., and McChesney, M.M. (1993). Air and fog deposition residues of four OP insecticides used on dormant orchards in the San Joaquin Valley. California. *Environ. Sci. Technol.* 27, 2236–2243. Copyright (1993) American Chemical Society.

it was observed for several pesticides and related chemicals by Glotfelty et al. (1986, 1987). It is generally assumed that the distribution of low-solubility organic solutes between air and water is described by Henry's constant (= air/water distribution coefficient) and that this distribution coefficient, calculated from the ratio of the vapor density (a direct function of vapor pressure) to the water solubility of the pure organic chemical (Eisenreich et al., 1981; Harder et al., 1980; Ligocki et al., 1985; Suntio et al., 1988), would prevail in fog.

Henry's law holds for vapor in equilibrium with the water phase. This equilibrium is achieved rapidly when there is pronounced surface contact between the two, such as apparently occurs in atmospheric moisture in clouds and raindrops (Majewski and Capel, 1995; Pankow et al., 1984).

Because enrichment was more pronounced for hydrophobic than hydrophilic pesticides, it was hypothesized that fog droplets contained solutes, such as dissolved or colloidal organic matter, which increased the solubility of hydrophobic chemicals over pure water. Surface active material had previously been reported in fog (Gill et al., 1983). Common atmospheric organics, such as α-pinene, n-hexanol, eugenol, and anethole, produced films with surfactant-like properties. Such films may exist in cloud droplets and snowflakes, in addition to fog droplets. These surface-active organics present at the air–water interface act to enhance the uptake of low-solubility organics into the aqueous phase (Chiou et al., 1986).

Although enrichment was reproducible in fog sampling conducted after that reported by Glotfelty et al. (1986, 1987), it was generally less marked in subsequent studies. Aqueous-phase enrichments for five chemicals ranged from 58 (diazinon), 42 (chlorpyrifos), 14 (parathion), and 1.4 (methidathion) in fog sampled in the San Joaquin Valley in 1986 (Glotfelty et al., 1990b, 1990c) (Table 7.2). There was large variability in EF for the same chemical measured in different fog events which had no obvious explanation. The liquid water content, which ranged from 0.024 to 0.080 g/m$^3$, total organic carbon content (38–55 mg/L), and pH (5.4–7) did not vary that much between samples and did not show correlational trends. Temperature effects were hypothesized to have played a small role. H generally increases with increasing temperature, and most of the calculated H constants were from data at 20°C or 25°C. Since the fog collections were made in atmospheres of 1.5°C–8.5°C, measured distributions (D) should be lower than those calculated from H, but only by factors of 2–4.

Seiber et al. (1993) found enrichment factors of only 5.7 (chlorpyrifos), 3.3 (diazinon), 2.7 (parathion), and 0.01 (methidathion). These magnitudes were considerably less than those of Glotfelty et al. (1986, 1987) and almost within the range predicted for temperature effects on distribution. Seiber et al. (1993) suggested the operation of variables, or sources of error, not yet considered or explained. It should be noted that Seiber et al. (1993) used a Teflon strand collector while the Glotfelty et al. (1986, 1987) collections were made with the rotating stainless steel screen device (Glotfelty et al., 1986).

Two primary hypotheses have emerged to explain the enrichment of organics in the aqueous phase of fogwater. (1) Organic solutes or nonfilterable colloids in the water phase enhance water solubility above that in pure water, or sorb organics so that more are bound in the fogwater phase than in pure water. (2) The air–water interface acts as a third phase, or compartment, in what had been assumed to be a simple 2-phase, or 2-compartment (viz. air and water) process. Capel et al. (1990) provided arguments in favor of the first hypothesis, essentially supporting the conclusions of Glotfelty et al. (1986, 1987). They argued that, on a mass basis, the "dissolved" organic fraction in fogwater is 10–100 times typical values reported for rain, lakes, and rivers. They also showed that the surface tension of fog is less than that of pure water, reflecting the presence of surface-active material. They assumed, then, that a portion of the fog droplet/air interface is covered with organic chemicals so that airborne chemicals confront an organic/air interface rather than a water/air interface. This would provide a mechanism for the concentration or enrichment of hydrophobic organic contaminants in the fog droplet.

Sagebiel and Seiber (1993) provided new evidence for the involvement of an interface in the enrichment process. They measured the air–water distribution of a number of phenolic products of incomplete combustion of wood in fogs collected in a winter-time residential setting in the Central Valley of California. The three most significant phenols, guaiacol, 4-methylguaiacol, and syringol, showed little to no enrichment. In plotting enrichment factor vs either octanol–water partition coefficient (Figure 7.6a) or water solubility (Figure 7.6b), it was clear that these wood smoke marker chemicals extended the range of solubilities represented by the pesticides studied by Glotfelty et al. (1987) to much higher water solubility (Figure 7.6b) or much lower octanol–water partition coefficient (Figure 7.6a), demonstrating that enrichment was limited to nonpolar organics in a regular fashion. This argued for a

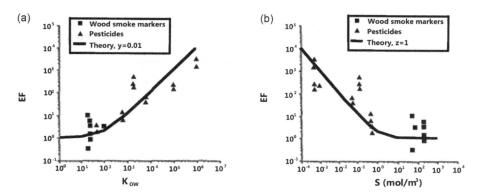

**FIGURE 7.6**

EF of pesticides and wood smoke markers plotted against octanol–water partition coefficient (a) and water solubility (b). (Reprinted with permission from Sagebiel, J.C., and Seiber, J.N. (1993). Studies on the occurrence and distribution of wood smoke marker compounds in foggy atmospheres. *Environ. Toxicol. Chem. Int. J.* 12, 813–822. Copyright (1993), Wiley.)

third phase in the air/fog system which might be made up of fine particulate matter, colloids, dissolved organic matter, or a surface film (first hypothesis). A theoretical treatment showed that this unknown sorptive phase could be described with the existing data for pesticides and wood smoke marker chemicals. The authors pointed out that they made no supposition regarding the composition of this third phase, but the concept of surface-active organics (first hypothesis) was one which was considered.

Goss (1994) summarized arguments in favor of the second hypothesis. He argued that the existence of an inseparable organic phase in which nonpolar chemicals accumulate provides only a partial explanation for the observations. Perona (1992), Valsaraj (1988), and Valsaraj et al. (1993) pointed to the importance of adsorption at the air–water interface as the missing process which had been overlooked. The very large specific area of small fog droplets provides an adsorptive phase which contributes to enrichment. Adsorption at this interface can be described by a partition coefficient relating the amount adsorbed to the equilibrium concentration in the air ($K_{ia}$) or to the equilibrium concentration in the water ($K_{iw}$). Valsaraj et al. (1993) found a correlation existed between $K_{iw}$ and $K_{ow}$ (octanol–water partition coefficient) for selected hydrophobic organics: $K_{iw}=3\times10^{-7}K_{ow}^{0.68}$ . They used this equation to predict the adsorption of pesticides at the air–water interface and predict the resulting enrichment in fog droplets. Goss (1994) developed a more complex equation, relating $K_{ia}$ and vapor pressure, which could be used in a similar way to that of Valsaraj et al. (1993). Applying this equation, and some temperature corrections to adjust the vapor pressure and fogwater collection, produced the results in Table 7.4. The calculated adsorption on the water surface led to a significant enrichment of chemicals in fogwater of the same magnitude as observed experimentally.

**TABLE 7.4**

Comparison of the Calculated (8 μm Droplet Diameter) and Measured Enrichment Factors

| Substance | $EF_{calc}$ | $EF_{lit}$ |
|---|---|---|
| Chlorpyrifos | 142 | 7–74 |
| | | 19–25 |
| Parathion | 39 | 4–29 |
| Methyl parathion | 11.7 | 0.7 |
| Malathion | 17.5 | 6 |
| Paraoxon | 1.4 | 2.1 |
| Diazinon | 273 | 6–160 |
| | | 30–50 |
| Atrazine | 1.8 | 0.05 |
| Alachlor | 1.76 | >4 |
| Pendimethalin | 2,476 | 1,500–3,200 |
| Fonofos | 30 | 3–5 |
| Guaiacol | 1.2 | 4.2 |
| | 1.2 | 3.3 |
| 4-Methylguaiacol | 1.4 | 3.0 |
| | 1.4 | 3.2 |
| Syringol | 10.0 | 9.6 |
| | 7.5 | 0.32 |

*Source:* Reprinted from Goss, K.-U. (1994). Predicting the enrichment of organic compounds in fog caused by adsorption on the water surface. *Atmos. Environ.* 28, 3513–3517. Copyright (1994), with permission from Elsevier.

Hoff et al. (1993) analyzed much of the same experimental data, but included new data on $K_{ia}$ and $K_{wa}$ determined by gas chromatographic retention times using water as a stationary phase. Their results supported that partitioning at the air–water interface can be appreciable and must be taken into account. Small droplets of water in air (fogs or clouds), small bubbles of air in water, and relatively dry low organic soils are examples. Correlations were derived for interface-air and interface-water coefficients for 44 polar and nonpolar organic chemicals.

## 7.6 Significance

The significance of pesticide residue occurrence in foggy atmospheres and fogwater warrants discussion. From a human health viewpoint, residues of individual chemicals in fogwater probably do not pose a significant risk

(Seiber et al., 1993). The higher concentrations recorded, 100 ng/m$^3$ in air and 100 µg/L in fogwater, may be expressed in human respiratory exposures as:

Interstitial Air: $0.1 \, \mu g/m^3 \times 15 \, m^3/day \times 1/70 \, kg = \sim 0.02 \, \mu g/kg/day$

Suspended Water: $100 \, \mu g/L \times 0.1 \times 10^{-3} L/m^3 \times 15 \, m^3/day \times 1/70 \, kg$

$$= \sim 0.002 \, \mu g/kg/day.$$

That is, for a 70-kg person breathing at 15 m$^3$/day, the exposure is 0.02 µg/ kg/day from breathing interstitial fog air, and 0.002 µg/kg/day for inhaling suspended water in the foggy air. For comparison, the acceptable daily intake (ADI) for parathion, perhaps the most toxic of the pesticides observed to date in fogwater, is 5 µg/kg/day, established by FAO/WHO based on an oral no-observed-effect level (NOEL) of 0.05 mg/kg/day for red blood cell acetylcholinesterase inhibition (Oudiz and Klein, 1988). The exposure from breathing foggy air is thus less than 1/1,000[th] of the ADI under these high exposure conditions. People do not breathe foggy air 24 hr/day, and fogs tend to be transient by their very nature. Thus it is hard to envisage conditions under which single chemical risks from breathing foggy air become significant. But the data on pesticides in fogwater show that mixtures of several chemicals, including several OPs and their oxons, may be present simultaneously so that it is not appropriate to dismiss fogwater altogether from the viewpoint of human health implications. More studies are needed, particularly with regard to mixtures and long-term exposures for people, including children and other sensitive subpopulations living in unusually fog-prone areas where pesticides are used.

Fogwater deposition to nontarget food crops represents an indirect exposure for humans. Turner (1989) found that fogwater deposition was a source of inadvertent residues to nontarget crops in California's Central Valley. But the residue contribution from this source is small, on the order of 0.01–0.1 ppm or less under worst case conditions. Thus the concern is not on human health impacts but rather on the legal question of what to do with food crops which receive low-level inadvertent residues of chemicals for which no tolerance has been established on crops where the contamination is found.

For wildlife dwelling in or very near orchards, the risks may be significantly higher than those for humans. Wildlife, such as birds, are exposed constantly to residues in the air. Birds located within the canopy of treated trees are exposed at much higher levels. Also, deposition of airborne residue to the feathers or fur may be significant because deposited residue may be ingested during preening. Birds may also contact residue through their feet or talons when roosting on a treated or exposed branch. Finally, all wildlife could ingest residue in their food. For red-tail hawks that frequent deciduous orchards in midwinter, inhalation, oral intake via preening, oral intake via food and water, and dermal intake through talons are all routes for exposure. It may all contribute to a residue exposure sufficiently high to produce biochemical and clinical signs of organophosphorus intoxication (Hooper et al., 1989; Wilson et al., 1991).

If one considers fog as simply one type of cloud, then the significance of fog- or cloudwater accumulation and transport of pesticides and other contaminants may assume a regional or global significance. The potential for regional or long-range transport and deposition of Valley-derived OPs used in dormant spraying is an example. Chlorpyrifos, diazinon, and parathion and their oxons were detected in air, rain, and snow samples collected in December–March in the Sierra Nevada Mountains southeast of the San Joaquin Valley orchard regions (Zabik and Seiber, 1993). Air concentrations at pg/m$^3$ levels were determined at elevations of 533 and 1,920 m in the mountains. Rainwater concentrations were at ppt (ng/L) levels at the mountain sites. Clearly, long-range transport of OPs can occur, but the indications are that removal processes (exchange with soil and plant surfaces, wet deposition, and chemical breakdown via oxon formation followed by hydrolysis) operating in the Valley itself may limit the buildup of OPs at nontarget sites outside the Valley.

Recently a number of pesticides and PCBs were detected in snow and some surface waters in the Sierra Nevada Mountains, apparently resulting from wet deposition of residues in clouds (McConnell et al., 1998). The significance of these aerially transported residues warrants further investigation (Seiber and Woodrow, 1998).

## 7.7 Conclusions

Fog represents an atmospheric medium in which pesticides may be degraded or undergo vapor–liquid distribution, and subsequent deposition. It probably does not alone present a direct health hazard to humans, but it will add to total exposure from ingestion of pesticides in food crops and drinking water, and residential exposure. Fogwater residues may have a major health impact on animals which reside in or around pesticide-treated areas, from deposition to their feathers or fur. More work is needed to define risks associated with pesticides, as well as other toxicants in foggy atmospheres.

Pesticide residues found in rainwater are frequently reported by the U.S. Geological Service as relevant to fog. Such residues are generally low compared with fogwater but result in runoff to lakes, streams, and other local waterways.

Text of this chapter was Reproduced with permission from Seiber, J.N. and Woodrow, J.E. (2000). Transport and Fate of Pesticides in Fog in California's Central Valley. In *Agrochemical Fate and Movement*, T.R. Steinheimer, L.J.

Ross, and T.D. Spittler, eds. (Washington, DC: American Chemical Society), pp. 323–346. Copyright (2000) American Chemical Society.

## References

Bidleman, T.F. (1988). Atmospheric processes. *Environ. Sci. Technol.* 22, 361–367.

Bidleman, T.F., and Olney, C.E. (1975). Long range transport of toxaphene insecticide in the atmosphere of the western North Atlantic. *Nature* 257, 475–477.

Capel, P.D., Gunde, R., Zuercher, F., and Giger, W. (1990). Carbon speciation and surface tension of fog. *Environ. Sci. Technol.* 24, 722–727.

Chernyak, S.M., Rice, C.P., and McConnell, L.L. (1996). Evidence of currently-used pesticides in air, ice, fog, seawater and surface microlayer in the Bering and Chukchi Seas. *Mar. Pollut. Bull.* 32, 410–419.

Chiou, C.T., Malcolm, R.L., Brinton, T.I., and Kile, D.E. (1986). Water solubility enhancement of some organic pollutants and pesticides by dissolved humic and fulvic acids. *Environ. Sci. Technol.* 20, 502–508.

Collett Jr, J.L., Daube Jr, B.C., Munger, J.W., and Hoffmann, M.R. (1990). A comparison of two cloudwater/fogwater collectors: The rotating arm collector and the caltech active strand cloudwater collector. *Atmospheric Environ. Part Gen. Top.* 24, 1685–1692.

Daube, Jr., B.C., Flagan, R.C., and Hoffman, M.R. (1987) Active cloudwater collector. United States Patent No. 4,697,462.

Eisenreich, S.J., Looney, B.B., and Thornton, J.D. (1981). Airborne organic contaminants in the Great Lakes ecosystem. *Environ. Sci. Technol.* 15, 30–38.

Gill, P., Graedel, T., and Weschler, C. (1983). Organic films on atmospheric aerosol particles, fog droplets, cloud droplets, raindrops, and snowflakes. *Rev. Geophys.* 21, 903–920.

Glotfelty, D.E., Majewski, M.S., and Seiber, J.N. (1990b). Distribution of several organophosphorus insecticides and their oxygen analogs in a foggy atmosphere. *Environ. Sci. Technol.* 24, 353–357.

Glotfelty, D.E., Schomburg, C.J., McChesney, M.M., Sagebiel, J.C., and Seiber, J.N. (1990c). Studies of the distribution, drift, and volatilization of diazinon resulting from spray application to a dormant peach orchard. *Chemosphere 21*, 1303–1314.

Glotfelty, D.E., Seiber, J.N., and Liljedahl, L.A. (1986). Pesticides and other organics in fog. In *Proceedings of EPA/APCA Symposium on Measurements of Toxic Air Pollutants: US EPA/Air Pollution Control Association.* Raleigh, NC: United States EPA, pp. 168–175.

Glotfelty, D.E., Seiber, J.N., and Liljedahl, A. (1987). Pesticides in fog. *Nature 325*, 602–605.

Glotfelty, D.E., Williams, G.H., Freeman, H.P., and Leech, M.M. (1990a). Regional atmospheric transport and deposition of pesticides in Maryland. *Long Range Transp. Pestic.* 199–221.

Goolsby, D.A., Thurman, E.M., Pomes, M.L., and Battaglin, W.A. (1994). Temporal and geographic distribution of herbicides in precipitation in the Midwest and Northeast United States, 1990–91. In *New Directions in Pesticide Research, Development, Management, and Policy: Proceedings of the Fourth National Conference on Pesticides*: Blacksburg, VA: Virginia Polytechnic Institute and State University, Virginia Water Resources Research Center, pp. 697–710.

Goss, K.-U. (1994). Predicting the enrichment of organic compounds in fog caused by adsorption on the water surface. *Atmos. Environ.* 28, 3513–3517.

Harder, H.W., Christensen, E.C., Matthews, J.R., and Bidleman, T.F. (1980). Rainfall input of toxaphene to a South Carolina estuary. *Estuaries* 3, 142–147.

Henderson, J., Yamamoto, J., Fry, D., Seiber, J., and Wilson, B. (1994). Oral and dermal toxicity of organophosphate pesticides in the domestic pigeon (Columba livia). *Bull. Environ. Contam. Toxicol.* 52, 633–640.

Hoff, J.T., Mackay, D., Gillham, R., and Shiu, W.Y. (1993). Partitioning of organic chemicals at the air-water interface in environmental systems. *Environ. Sci. Technol.* 27, 2174–2180.

Hooper, M.J., Detrich, P.J., Weisskopf, C.P., and Wilson, B.W. (1989). Organophosphorus insecticide exposure in hawks inhabiting orchards during winter dormant-spraying. *Bull. Environ. Contam. Toxicol.* 42, 651–659.

Jacob, D.J., Wang, R.F.T., and Flagan, R.C. (1984). Fogwater collector design and characterization. *Environ. Sci. Technol.* 18, 827–833.

Kurtz, D.A. (1990). *Long Range Transport of Pesticides.* Chelsea, MI: CRC Press.

Lewis, R.G., and Lee, R.E. (1976). *Air Pollution from Pesticide and Agricultural Processes.* Cleveland, OH: CRC Press.

Ligocki, M.P., Leuenberger, C., and Pankow, J.F. (1985). Trace organic compounds in rain—III. Particle scavenging of neutral organic compounds. *Atmospheric Environ.* 1967 19, 1619–1626.

Majewski, M.S., and Capel, P.D. (1995). *Pesticides in the Atmosphere.* Chelsea, MI: Ann Arbor Press.

Mallant, R.K., and Kos, G.P. (1990). An optical device for the detection of clouds and fog. *Aerosol Sci. Technol.* 13, 196–202.

McConnell, L.L., LeNoir, J.S., Datta, S., and Seiber, J.N. (1998). Wet deposition of current-use pesticides in the Sierra Nevada mountain range, California, USA. *Environ. Toxicol. Chem. Int. J.* 17, 1908–1916.

Oudiz, D., and Klein, A. (1988). *Evaluation of Ethyl Parathion as a Toxic Air Contaminant.* Sacramento, CA: California Department of Food and Agriculture, Environmental Monitoring and Pest Management (Report No. EH-88-5).

Pankow, J.F., Isabelle, L.M., and Asher, W.E. (1984). Trace organic compounds in rain. 1. Sampler design and analysis by adsorption/thermal desorption (ATD). *Environ. Sci. Technol.* 18, 310–318.

Perona, M.J. (1992). The solubility of hydrophobic compounds in aqueous droplets. *Atmos. Environ. Part Gen. Top.* 26, 2549–2553.

Rice, C.P. (1996). *Pesticides in Fogwater.* Pestic. Outlook U. K.

Sagebiel, J.C., and Seiber, J.N. (1993). Studies on the occurrence and distribution of wood smoke marker compounds in foggy atmospheres. *Environ. Toxicol. Chem. Int. J.* 12, 813–822.

Schomburg, C.J., Glotfelty, D.E., and Seiber, J.N. (1991). Pesticide occurrence and distribution in fog collected near Monterey, California. *Environ. Sci. Technol.* 25, 155–160.

Seiber, J.N., Glotfelty, D.E., Lucas, A.D., McChesney, M.M., Sagebiel, J.C., and Wehner, T.A. (1990). A multiresidue method by high performance liquid chromatography-based fractionation and gas chromatographic determination of trace levels of pesticides in air and water. *Arch. Environ. Contam. Toxicol.* 19, 583–592.

Seiber, J.N., Wilson, B.W., and McChesney, M.M. (1993). Air and fog deposition residues of four organophosphate insecticides used on dormant orchards in the San Joaquin Valley, California. *Environ. Sci. Technol.* 27, 2236–2243.

Seiber, J.N., and Woodrow, J.E. (1995). Origin and fate of pesticides in air. In *Proceedings of Eighth International Congress of Pesticide Chemistry: Options*. Washington, DC: American Chemical Society, pp. 157–172.

Seiber, J.N., and Woodrow, J.E. (1998). Air transport of pesticides. *Rev. Toxicol.* 2, 287–294.

Spear, R.C., Popendorf, W.J., Leffingwell, J.T., and Jenkins, D. (1975). Parathion residues on citrus foliage. Decay and composition as related to worker hazard. *J. Agric. Food Chem.* 23, 808–810.

Steinheimer, T.R., Ross, L.J., and Spittler, T.D. (2000). ACS SYMPOSIUM SERIES 751-Agrochemical Fate and Movement-Perspective and Scale of Study. In *ACS Symposium Series*, Washington, DC: American Chemical Society, [1974], p.

Suntio, L., Shiu, W., Mackay, D., Seiber, J., and Glotfelty, D. (1988). Critical review of Henry's law constants for pesticides. In *Reviews of Environmental Contamination and Toxicology*, (pp. 1–59). New York, NY: Springer.

Thomas, T.C., and Seiber, J.N. (1974). Chromosorb 102, an efficient medium for trapping pesticides from air. *Bull. Environ. Contam. Toxicol.* 12, 17–25.

Turner, B. (1989). A field study of fog and dry deposition as sources of inadvertent pesticide residues on row crops. Environmental Hazards Assessment Program. State of California, Department of Food and Agriculture. Sacramento, CA. EH-89-11, 42p + appendices.

Valsaraj, K.T. (1988). On the physico-chemical aspects of partitioning of non-polar hydrophobic organics at the air-water interface. *Chemosphere* 17, 875–887.

Valsaraj, K.T., Thoma, G.J., Reible, D.D., and Thibodeaux, L.J. (1993). On the enrichment of hydrophobic organic compounds in fog droplets. *Atmos. Environ. Part Gen. Top.* 27, 203–210.

Wehner, T.A., Woodrow, J.E., Kim, Y.-H., and Seiber, J.N. (1984). Multiresidue analysis of trace organic pesticides in air. In L.H. Keith (Ed.), *Identification and Analysis of Organic Pollutants in Air*. Woburn, MA: Butterworth Publishers, pp. 273–290.

Wilson, B.W., Hooper, M.J., Littrell, E.E., Detrich, P.J., Hansen, M.E., Weisskopf, C.P., and Seiber, J.N. (1991). Orchard dormant sprays and exposure of red-tailed hawks to organophosphates. *Bull. Environ. Contam. Toxicol.* 47, 717–724.

Woodrow, J., Seiber, J., Crosby, D., Moilanen, K., Soderquist, C., and Mourer, C. (1977). Airborne and surface residues of parathion and its conversion products in a treated plum orchard environment. *Arch. Environ. Contam. Toxicol.* 6, 175–191.

Woodrow, J.E., Crosby, D.G., and Seiber, J.N. (1983). Vapor-phase photochemistry of pesticides. In F.A. Gunther, and J.D. Gunther, (Eds.), *Residue Reviews*. New York: Springer.

Yvon-Lewis, S.A., and Butler, J.H. (1997). The potential effect of oceanic biological degradation on the lifetime of atmospheric CH3Br. *Geophys. Res. Lett.* 24, 1227–1230.

Zabik, J.M., and Seiber, J.N. (1993). Atmospheric transport of organophosphate pesticides from California's Central Valley to the Sierra Nevada Mountains. *J. Environ. Qual.* 22, 80–90.

# 8

# *Fumigants*

## 8.1 Introduction

Millions of pounds of fumigants are applied each year to control pests at all stages of food production. Fumigants are important for production of strawberries, other specialty fruits, and grains. Applied preplant and subsequently vented, they control a wide array of soil-borne pests including nematodes, insects, and weeds. There is a growing need for fumigants use in stored products before they are transported and processed into a number of food products. Spices, grains, and nuts routinely receive fumigant applications. Fumigants are volatile low-molecular-weight chemicals that can diffuse into and out commodities or soil (Figure 8.1). Once their job of controlling pests is complete, they can be vented. Generally, the threated commodity can then be handle safely and eventually consumed free of pests (and pesticide residues). And fumigated fields, once vented, can grow a new crop of seedlings. California leads the United States in producing these commodities, both for the rest of the United States and for the export market. Hence, it is the largest user of fumigants (Table 8.1).

On the negative side, fumigants are generally hard to confine and become unwanted air contaminants after they do their job of killing pathogens and microorganisms that cause spoilage. Fumigants are also toxic to people, animals, and nontarget species, so there are many restrictions on use and undue exposures (Table 8.2). Among the fumigants listed, chloropicrin is the most toxic (Lee et al., 2002). Methyl bromide (MeBr) is of low acute toxicity but is capable of producing mutagenic or carcinogenic effects upon repeated exposures in mammals including humans. Therefore, to protect nontarget organisms, special precautions are needed when fumigants are used (Randy Segawa, personal communication, CDPR, April 19, 2018).

Fumigants are mobile compounds in the environment and warrant exceptional safeguards in terms of application technology, minimizing worker exposure, and preventing movement to air and groundwater. They pose risks to applicators and field workers, and to a lesser extent, to those who live in the vicinity of fumigant operations. In the past, regulatory agencies have banned agricultural chemicals that are too mobile and toxic to guarantee

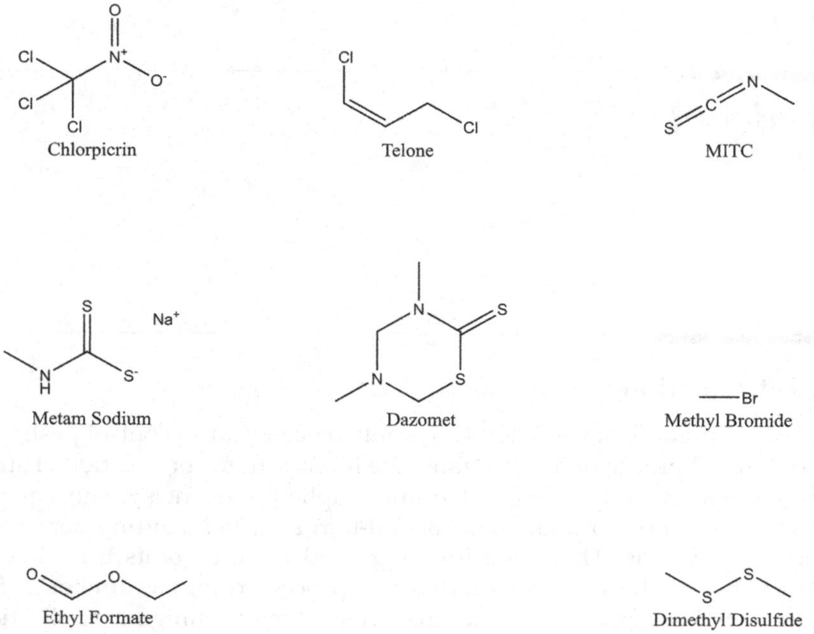

**FIGURE 8.1**

Structures of top fumigants. MITC is formed in soil from metam sodium, a precursor. Ethyl formate is a replacement biofumigant and wild mustard mustard is an effective natural nematicide—both show promise as alternatives to synthetic commercial fumigants. Dazomet and Metam both release the MITC fumigant in moist soil.

**TABLE 8.1**

Fumigants in the Top 25 Pesticides Used in California in 2018, Ranked by Pounds Applied

| Rank | Chemical | Pounds |
|---|---|---|
| 3 | 1,3-Dichloropropene (1,3-D; Telone) | 12,569,270 |
| 6 | Chloropicrin | 7,436,425 |
| 10 | Metam-sodium[a] | 3,765,705 |
| 20 | MeBr | 1,682,989 |

*Source:* California Department of Pesticide Regulation (2018).

[a] Metam-sodium degrades to the active ingredient MITC.

safety to people and the environment, or at least restricted their use. The fumigant class has been hit hard in this regard.

Many fumigants are water soluble and as a result can leach into groundwater and ultimately contaminate drinking water. This was the case with ethylene dibromide (EDB) and dibromochloropropane (DBCP); these stable, mobile, and potential carcinogens were found in groundwater and thus

**TABLE 8.2**

Fumigant Inhalation Reference Concentration (RfC in µg/m³)

| Fumigant | Acute (1–24 hour) | Subchronic (≥15 days) | Chronic (>1 year) |
|---|---|---|---|
| MeBr | 210 | 2 (6 weeks) | 5 |
| Chloropicrin | 29 | 1 | 1 |
| 1,3-D | 55 | 14 | 9 |
| MITC | 66 (1–8 hour) | 3 | 0.3 |
| Carbon disulfide (CS₂) | – | – | 700[a] |

*Source:* Lee et al. (2002).
*Note:* Chloropicrin is the most toxic.
[a] IRIS 1987, baseline: 55,100 µg/m³.

banned in 1979 (Babich et al., 1981; United States Environmental Protection Agency, 1992). MeBr, as discussed below, is rapidly released to the air after application to soil. Mitigation efforts including tarping and deep injection slow its release, though do not completely prevent it. Methyl isothiocyanate (MITC), 1,3-dichloropropene (1,3-D), chloropicrin, ethyl formate, and other chemicals are among the current replacements. Of the remaining fumigants, virtually all—MeBr, 1,3-D (Telone), and ethylene oxide—have been threatened with severe limitations. Fortunately, there may be time to learn more about them so that a ban is not necessary. To avoid a ban, we must be able to control exposures as well as air and groundwater contamination.

## 8.2 Methyl Bromide

MeBr, first used as a fumigant in the 1940s, has long been regarded as the fumigant against which others are compared. It is effective, leaves no residue, and is easy to produce and apply. Its low cost is a result of its heavy use in industry as a methylating agent to make methyl esters from carboxylic acids and methyl ethers from alcohols and other steps in synthesis. MeBr can control a whole spectrum of pests and is relatively low in toxicity to people and wildlife. It is removed after use by volatilization and other dissipation pathways such that it leaves no residue in food crops.

MeBr is best applied to soil using a fumigator, which delivers MeBr gas from cylinders chiseled into soil (Figure 8.2). The field is covered with a plastic tarp to confine the fumigant long enough to kill organisms in soil in several hours or days. It is best done by contracting with a firm with experience such as Tri-Cal; application depth, rate and type of fumigant and tarp makeup are among the variables. Fumigation of commodities like almonds and spices; perishables like blueberries; lumber; and other materials which require special precautions are done in chambers or bins following instructions from

**FIGURE 8.2**
A fumigator applying MeBr to soil preplant. Note the plastic tarp to cover soil after application.

the U.S. Department of Agriculture (USDA) Agricultural Research Services (ARS) and Animal and Plant Health Inspection Service (APHIS), and other federal and state authorities.

MeBr is transported to the air where it is estimated to be stable for 0.8 years. MeBr is recalcitrant to photodegradation at the wavelengths of sunlight prevalent in the troposphere. Instead, it is thought to be degraded slowly by hydroxyl radicals (Equation 8.1) (Honaganahalli and Seiber, 1996; NASA Evaluation No. 10).

$$CH_3Br + OH^\bullet \rightarrow CH_2Br^\bullet + H_2O \qquad (8.1)$$

MeBr moves to the stratosphere where the UV radiation is sufficient to photodissociate MeBr to release Br atoms. Bromine and chlorine react with and deplete ozone, as illustrated in Equation 8.2 (Honaganahalli and Seiber, 1996; Yung et al., 1980).

$$Br + O_3 \rightarrow BrO + O_2$$
$$Cl + O_3 \rightarrow ClO + O_2 \qquad (8.2)$$
$$BrO + ClO \rightarrow Br + Cl + O_2$$

$$\text{Net: } 2\,O_3 \rightarrow 3\,O_2$$

Bromine catalysis is most efficient in the lower stratosphere where the ozone concentration is largest. Another possible catalytic cycle is that between BrO and $HO_2$ (Equation 8.3) (Honaganahalli and Seiber, 1996; Poulet et al., 1992).

$$Br + O_3 \rightarrow BrO + O_2$$
$$BrO + HO_2 \rightarrow HOBr + O_2$$
$$HOBr + hv \rightarrow OH + Br \qquad (8.3)$$
$$OH + O_3 \rightarrow HO_2 + O_2$$

$$\text{Net: } 2\,O_3 \rightarrow 3\,O_2$$

Most nations have agreed to phase out ozone depleters like MeBr per the Montreal Protocol agreements. However, the phaseout has exemptions which the United States has applied for since 2004. The exemptions are based on the idea that fumigants are vital agents in the production of many food crops, particularly high-value crops such as strawberries and grapes, which are susceptible to nematodes and other soil-borne pests. They are also used—even required—before and during shipment of fruits, grains, and spices for export (Phillips et al., 2012; Walse, USDA ARS). These exemptions contend there are no suitable alternatives, thus some MeBr use continues even though it is still subject to phase out (see Table 8.1). The use has declined over time partly due to threat of phaseout and partly because growers have implemented alternatives to some MeBr uses. These alternatives include chloropicrin and 1,3-D. Nonchemical alternatives have also been developed such as soil solarization and use of nematode-resistant crop plants.

MeBr is predominantly a naturally occurring compound; anthropogenic (synthetic) sources represent only around 25% (±10%) of total emissions. Furthermore, oceans and soil were identified as sinks, which significantly decreased its ozone depletion potential from 0.7 to about 0.45 and overall lifetime from 2.0 to between 0.8 and 1.0 years (Honaganahalli and Seiber, 1996). (Note that $CO_2$ has an ozone depletion of 1.0.) Further, MeBr has tested mostly negative as a carcinogen. Stringent fumigation rules, at least in the United States, have been effective in reducing exposure. In light of this information, MeBr may be a better fumigant than some of its replacements. Instead of an outright ban, better management practices may be sufficient for its continued use as a fumigant (Honaganahalli and Seiber, 1996).

Recent controversy over the potential role of MeBr in damaging the ozone layer has spurred interest in increasing our understanding of the transformation and movement of this fumigant after it is applied to soil. Our research indicates MeBr is rapidly volatilized from fumigated soil (usually within the first 24 hours) and volatility significantly increases with temperature ($35°C > 25°C > 15°C$) and moisture ($0.03\ bar \geq 0.3\ bar > 1\ bar > 3\ bar$). Degradation of MeBr, measured by production of bromide ion ($Br^-$), was also directly related to temperature and moisture. Undisturbed soil column studies indicated that MeBr rapidly volatilized (>50% of the MeBr flux occurred in 48 hours) but did not leach into subsurface soil in that amount of time. Residual MeBr was degraded eventually in the soil column, evident by measurable concentrations of $Br^-$ in the leachate water. In field studies, MeBr also volatilized rapidly from soil, but a significant portion of the MeBr was degraded (30% after 2 days). These studies provide pertinent information for assessing the fate of MeBr in soil, which should lead to more informed decisions regulating its use. (Reprinted with permission from Anderson et al. (1996). Fate of Methyl Bromide in Fumigated Soils. In Fumigants, (American Chemical Society), pp. 42–52.). Copyright (1996) American Chemical Society.)

## 8.3 Emissions of MeBr from Agricultural Fields: Rate Estimates and Methods of Reduction

Numerous scientists have studied the environmental fate and transport of MeBr; searched for replacement chemicals and/or nonchemical alternatives; and developed new methodology which improves containment of MeBr (or any alternative fumigant) to the treatment zone, while maintaining adequate pest control. Research on the environmental fate and transport of MeBr and other fumigants follows the general U.S. Environmental Protection Agency (EPA) paradigm for exposure assessment, risk assessment, and risk management (Figure 8.3).

There are several methods for measuring emission rates (Majewski et al., 1990). These methods take into account measured gradients of wind speed and concentrations of emitted pesticide to calculate field-wide emission rates (Figure 8.4). The appropriate flux field design is chosen based on the method to be used to calculate from field data (Figure 8.5): with the aerodynamic (vertical) flux method, multitiered masts are placed in the center of the field and with the integrated horizontal flux method, the multitiered masts are placed at the downwind edge of treated fields. A database of measured emission rates can be used to estimate emission rates for new application

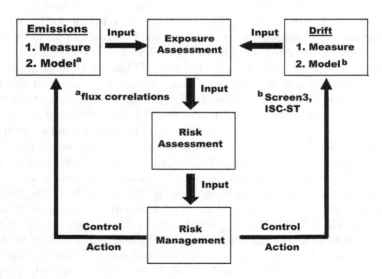

**FIGURE 8.3**
General EPA paradigm for exposure assessment, risk assessment, and risk management that was applied to fumigants for the field application. (Reprinted (adapted) with permission from Woodrow, J.E., Seiber, J.N., and Miller, G.C. (2011). Correlation to estimate emission rates for soil-applied fumigants. *J. Agri. Food Chem.* 59, 939–943. Copyright (2011) American Chemical Society.)

## Methods for Determining Emission Losses of Fumigants

Aerodynamic Gradient Method
$$ER = k^2 \Delta c \Delta u / \Phi m \Phi p [Ln(z_2/z_1)]^2)$$

Integrated Horizontal Flux
$$ER = (1/X) \int c_i u_i dz$$

Flux Chambers
$$ER = (V/A)(\Delta c/\Delta t)$$

Back Calculation
Downwind conc. + ISC-ST model

**FIGURE 8.4**
Methods for determining emission losses of fumigants in the field. For more on field chambers, the reader is referred to Yates (2006).

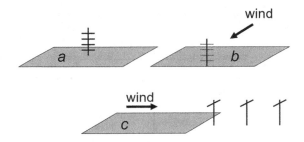

**FIGURE 8.5**
Flux field designs showing options for placing air samplers for (a) the aerodynamic flux method, (b) the integrated horizontal flux method, and (c) dispersion measurements.

scenarios through numeric techniques that correlate pesticide physicochemical properties with measured emissions (Woodrow et al., 2011).

To measure dispersion, single tiered masts are placed in downwind arrays (Figure 8.5). However, these laborious dispersion assays can be replaced with Gaussian plume dispersion models that use the measured emissions rate and weather data to approximate downwind concentrations for exposure/risk assessment (see Chapter 5) (Honaganahalli and Seiber, 2000; Woodrow et al., 1997). These computer-based Gaussian plume dispersion models can also be used to back-calculate a flux term when the downwind concentrations are known, for comparison with measured flux (Seiber et al., 1996).

### 8.3.1 Flux from a Single Fallow Field

To illustrate, a field study was conducted of MeBr volatilization flux, dispersion, and atmospheric fate in Monterey County, California, in 1994 (Seiber et al., 1996). Flux, vertical profile, and single-height air concentrations were measured at four different locations extending from the center of the treated field to nearly 0.8 km downwind (Figure 8.6). MeBr trapped on charcoal

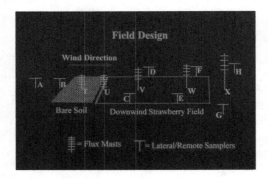

**FIGURE 8.6**
Field design for measuring downwind concentrations of MeBr from a fallow downwind field.

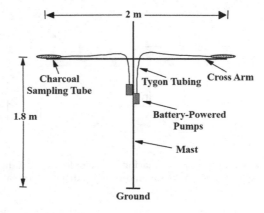

**FIGURE 8.7**
Detail of the air sampling masts for sampling MeBr in air prior to analysis by headspace-GC. (Reprinted (adapted) with permission from (Woodrow, J.E., LePage, J.T., Miller, G.C., and Hebert, V.R. (2014). Determination of MIC in outdoor residential air near metam-sodium soil fumigations. *J. Agri. Food Chem.* 62, 8921–8927.). Copyright (2014) American Chemical Society.)

adsorbent was subsequently extracted and subjected to gas chromatography (GC) analysis (see Chapter 6) (Figure 8.7). Air concentrations of MeBr measured above a bare fumigated field and downwind from the field along with meteorological data were used to calculate vertical (aerodynamic) and horizontal flux, respectively. Mid-field vertical flux and average horizontal flux measured on the downwind edge of the field were all but identical (~99%). Vertical air concentration and flux profiles farther downwind showed depletion of MeBr near the surface of an adjoining mature strawberry field (Figure 8.6) and adsorbed MeBr was found in the surface soil of the same field (preliminary evidence that soil is a sink for MeBr) (Seiber et al., 1996). Another objective was to compare downwind air concentration data to concentrations predicted by a dispersion model (Industrial Source

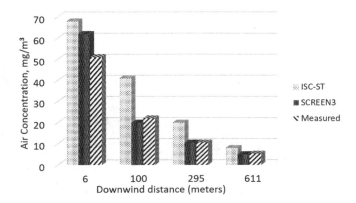

**FIGURE 8.8**
MeBr air concentrations measured at different distances from downwind of field.

Complex-Short Term II [ISC-STII]) in order to assess the value of this model for use in exposure assessment for agricultural workers and downwind residents (Seiber et al., 1996). As expected, measured air concentrations of MeBr fell off exponentially with downwind distance (Figure 8.8). MeBr air concentrations measured downwind of the treated field source compared well with those predicted by the ISC-STII model, although measured dispersion values were less than modeled dispersion values by a factor of ~2. Later, application of the SCREEN-3 model allowed for better approximation of actual measured values.

Section 8.3.1: (Reprinted (adapted) with permission from Seiber et al. (1996). Flux, Dispersion Characteristics, and Sinks for Airborne Methyl Bromide Downwind of a Treated Agricultural Field. In Fumigants, pp. 154–177. Washington, D.C.: American Chemical Society. Copyright (1996) American Chemical Society.)

## 8.3.2 Flux from Multiple Sources

Estimates of downwind concentrations of pollutants from multiple agricultural sources simultaneously were addressed in a field study conducted in September 1995 (Honaganahalli and Seiber, 2000). The ambient atmospheric concentrations of MeBr in the Salinas Valley, California, were measured following application of MeBr to multiple agricultural fields (Figure 8.9). To catch the spreading plume, air samples of MeBr were measured at 11 sites located on the adjacent mountains, valley floor, and at the Pacific Ocean coast each night and day over a 4-day period. The laborious field experiment could then be adjusted for weather, topography, application rate and method, and tarp composition using computational models provided by EPA (ISCST) and California EPA (CALPUFF). MeBr exposure for the population of the city of Salinas was calculated based on the measured ambient concentrations and

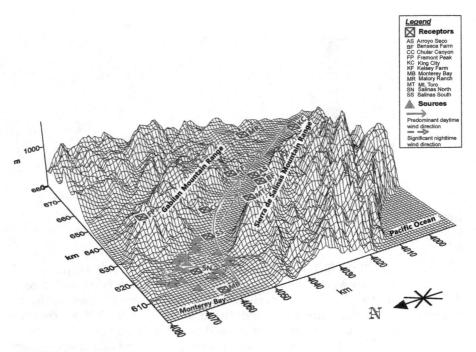

**FIGURE 8.9**
Predicted movement of MeBr plumes based on wind in the Salinas Valley airshed. Several fields located near the Pacific Coast were fumigated with MeBr at different times, which was carried by prevailing winds. This shows an approach to estimating airshed MeBr concentrations when there are several fields as sources in the same general location. It also shows how the emission equation works even for several fields treated separately with MeBr. (Reprinted from Honaganahalli, P.S., and Seiber, J.N. (2000). Measured and predicted airshed concentrations of MeBr in an agricultural valley and applications to exposure assessment. *Atmos. Environ.* 34, 3511–3523. Copyright (2000), with permission from Elsevier.)

compared with the current benchmark used by U.S. EPA and the California Department of Pesticide Regulation (CDPR) for acceptable human health risk.

### 8.3.2.1 Flux from Multiple Fields: Ambient Concentrations Downwind

MeBr in air collected at 11 sites was trapped on charcoal adsorbent (Figure 8.7), extracted, and subjected to GC analysis (Honaganahalli and Seiber, 2000). Measured concentrations of MeBr are reported in Table 8.3.

The concentration distribution in the valley suggests that wind strength and direction differences at day and night strongly influence ambient MeBr concentrations at downwind locations (Figures 8.9 and 8.10). The rapid winds during the day in the Salinas Valley transported MeBr to the southern end of the valley, mainly along the Gabilan slopes, leading to significant lateral and forward dispersion of the MeBr plume. Highest daytime concentrations

**TABLE 8.3**

A Summary of Measured Ambient Air Concentrations ($\mu$g/m$^3$) of MeBr in Salinas Valley, CA, on September 1–4, 1995, at all locations

| Day | MB | SN | SS | AS | KC | FP | CC | KF | MR | MT | BF |
|---|---|---|---|---|---|---|---|---|---|---|---|
| 9/1 | ND[a] | 0.654 | FE[b] | 0.974 | 0.54 | 1.23 | 2.12 | 0.777 | FE | 0.041 | 0.658 |
| 9/2 | ND | 0.3 | 0.751 | 0.656 | <LOD[c] | 1.21 | 3.55 | ND | ND | 0.156 | ND |
| 9/3 | 1.37 | 1.85 | 0.881 | 0.599 | 0.445 | 1.05 | 3.16 | 1.94 | 0.765 | 0.064 | 0.819 |
| 9/4 | FE | 0.717 | 0.807 | 0.651 | 0.319 | 0.652 | 1.13 | 0.98 | 0.604 | 0.148 | 0.623 |
| Avg | – | 0.881 | 0.813 | 0.72 | 0.435 | 1.04 | 2.49 | 1.23 | 0.684 | 0.102 | 0.7 |
| **Night** | | | | | | | | | | | |
| 9/1 | ND | FE | 2.22 | 0.294 | 0.17 | 0.231 | 0.873 | 1.09 | 0.708 | <LOD[c] | 0.178 |
| 9/2 | 1.05 | 8.37 | 3.43 | 2.15 | 0.316 | ND | 2.01 | ND | ND | 0.407 | ND |
| 9/3 | 0.25 | 3.36 | 8.98 | 1.16 | 0.551 | 1.13 | 0.773 | 1.7 | 2.14 | 0.117 | 2.16 |
| Avg | 0.65 | 5.86 | 4.88 | 1.2 | 0.346 | 0.68 | 1.22 | 1.4 | 1.42 | 0.202 | 1.17 |

*Source:* Reprinted from Honaganahalli, P.S., and Seiber, J.N. (2000). Measured and predicted airshed concentrations of methyl bromide in an agricultural valley and applications to exposure assessment. *Atmos. Environ.* 34, 3511–3523. Copyright (2000), with permission from Elsevier.

[a] LOD, Limit of detection = 48 ng/m$^3$.
[b] ND, No determination.
[c] FE, Field error sample lost.

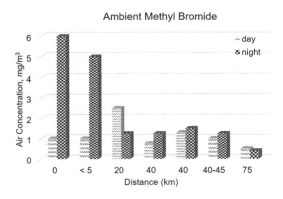

**FIGURE 8.10**
MeBr air concentrations day vs night (measured). Wind was calmer at night. (Data from Honaganahalli and Seiber 2000.)

were on the windward face of the Gabilan mountain range (CC, 20 km, 3.55 $\mu$g/m$^3$; KF, 40 km, 1.94 $\mu$g/m$^3$). These winds enabled the farthest receptor (KC, ~70 km) to record appreciable concentrations (0.55 $\mu$g/m$^3$), even though there were no nearby sources, compared to 0.654 $\mu$g/m$^3$ at SN (0 km) (Honaganahalli and Seiber, 2000).

In the northern region of the valley at night, a relatively calmer (than daytime) land breeze blows westward to the Pacific Ocean or southward along the valley floor (Figure 8.9, dashed arrows) (Honaganahalli and Seiber, 2000). These calmer nighttime winds resulted in the highest measured concentrations of MeBr to be around the city of Salinas at night (8.98 and 8.37 µg/m³ at SS and SN). In contrast, daytime levels of MeBr did not exceed (0.881 and 1.85 µg/m³ at SS and SN). Similar average concentrations during the day (0.435 µg/m³) and at night (0.346 µg/m³) at KC suggest a declining diurnal influence over the southern part of the valley.

On leeward (east-facing) slopes of the Sierra de Salinas range, both during daytime and nighttime consistently, lower concentrations were recorded at higher elevations (e.g., MT), and higher concentrations at lower levels (e.g., BF). This could be due to the funneling of the air mass containing MeBr toward the southern end of the valley and not vertically. At MB, the presumed upwind site, concentrations were observed mostly when the site was under the influence of a land breeze at night. MeBr, which was transported over the ocean at night, may have returned the following day with the sea breeze. This may explain MeBr presence at the MB receptor during one daytime session.

### 8.3.2.2 Flux from Multiple Fields: ISCST3 Model Performance

The EPA ISCST dispersion model is a general model that can be used for flat or rolling terrain. It is built on a straight line, steady-state Gaussian plume equation. It is best used for transport distances less than 50 km and works for a variety of averaging times. It was adapted to agricultural fields similar to the aforementioned bare strawberry field fumigated with MeBr (Seiber et al., 1996).

The major input terms are (1) MeBr applications at the source, (2) source flux emission rates, and (3) meteorological data (wind speed, direction, surface air temperature, stability class [fluctuations in wind], and urban/rural mixing height). The output is average MeBr concentrations for each location during the period indicated.

MeBr sources (fields treated with MeBr) within the Salinas Valley were identified from the Monterey County Agricultural Commissioner's Annual Pesticide Use Report database (MCAPUR) (Honaganahalli and Seiber, 2000). Flux was not measured in individual fields; this source term was adapted from two estimates from field studies obtained under similar but not identical conditions (Seiber et al., 1996; Yates et al., 1996).

ISCST3 simulations used meteorological data from the closest California Irrigation Management Information Systems (CIMIS) station (Honaganahalli and Seiber, 2000). "The ISCST3 model requires meteorology that is representative over the entire modeling domain. Because Salinas Valley has such complex wind patterns, meteorological data obtained from any single weather station in the valley will not represent the meteorology prevailing over the entire valley." In the north, near the receptors, nighttime breezes were from the mountain slopes toward the Pacific Ocean. But in the middle and southern regions of the valley, southerly winds prevailed. The problem was compensated by running the model twice: once using the meteorology near the source (north) and again using the meteorology at the receptors (south). More detailed meteorological considerations, such as rotation and stability index, are described in the article (Honaganahalli and Seiber, 2000).

Concentration patterns, highs and lows, and trends at a receptor exhibited diurnal differences somewhat similar to measured concentration patterns. Just as in the measured data, the diurnal meteorological patterns prevailing in the valley strongly influenced the distribution of MeBr in the valley airshed. In agreement with measured concentrations during daytime, the ISCST3 model predicted higher concentrations on the windward slopes of the Gabilan range than on the valley floor (i.e., CC); at night, the model predicted higher concentrations on the valley floor in the northern part (SN, SS).

Statistical analysis showed moderate correlation considering 24-hour time-weighted average concentrations ($R^2 \approx 0.70$) and nighttime concentrations ($R^2 \approx 0.7$), but poor daytime correlation ($R^2 \approx 0.2$). We suspected the reasons for this poor daytime performance of the model was due to the relatively unstable atmospheric conditions throughout the valley, and that using different degrees of "rotation" for daytime and nighttime would result in higher daytime correlations. For additional details regarding the analysis of MeBr field experiment, see Honaganahalli and Seiber (2000).

### 8.3.2.3 *Flux from Multiple Fields: CALPUFF Model Performance*

The CALPUFF model, using 3D meteorology provided by CALMET, was successful both physically and statistically in predicting the nighttime concentrations but was less successful in predicting concentrations at all stations during daytime. During daytime, the CALPUFF model predicted concentrations only at four sites in the northern part, two of which were on the valley floor (SN and SS). At all other sites, the model predicted zero or insignificant concentrations. During nighttime, CALPUFF predicted concentrations at all receptors.

The correlation between measured and CALPUFF-predicted 24-hour time-weighted average concentrations ($R^2 = 0.55$ and $0.82$ for Seiber et al.'s (1996) and Yates et al.'s (1996) source inputs, respectively) and nighttime concentrations ($R^2 = 0.89$ and $0.90$) suggest that the CALPUFF model was moderately successful in predicting the measured concentrations. For daytime, the correlation between measured and model predicted values was poor ($R^2 = 0.22$ and $0.24$). This mixed result points at the inherent limitations of the CALMET model to provide precise turbulence information needed for accurate estimation of dispersion parameters in a highly complex terrain.

For the CALMET model, topographic information of high resolution, which could have resulted in estimation of better wind flow patterns, was not available. Also, all surface meteorological stations in the domain were clustered in and around the city of Salinas. Among the available meteorological stations, all required elements were not available at all stations. For all of these reasons, CALMET was unable to resolve, especially in the daytime, the fine structure of the meteorology prevailing in the modeling domain explaining the lower correlation observed with CALMET/CALPUFF.

The lack of representative meteorology, variability in flux estimates, source–receptor spatial relationship, and inherent uncertainty in measurements are among the reasons for moderate correlation between measured and modeled concentrations, especially for daytimes. With refined source estimates and better meteorological data, improved model predictions can be expected. We also note that this result came from a single field experiment. Ideally it should be replicated to generate average values.

### 8.3.2.4 Flux from Multiple Fields: Applications to Exposure Calculations

Nearly 80% of the MeBr applications in the Salinas Valley occurred in the vicinity of the city of Salinas potentially exposing the population of the city of Salinas to MeBr. MeBr has been shown to cause developmental toxicity in animal studies and neurotoxicity in all species studied including humans (Lim, 1992). The U.S. EPA has established a chronic exposure reference concentration of 5 $\mu g/m^3$. The CDPR has established a 24-hour exposure limit of 816 $\mu g/m^3$ (210 ppb) (Nelson, 1992) and ensures compliance by enforcing buffer zones between the application area and places where people live or conduct activities for a significant period of time (e.g., schools).

### 8.3.2.5 Flux from Multiple Fields: Exposure Assessment

The margin of exposure (MOE) is the ratio of the no observed adverse effect level (NOAEL) ($mg/m^3$) from experimental animal studies to the human exposure ($mg/m^3$) concentration. The MOEs for acute exposure can be estimated from the highest airborne MeBr concentration measured at each location during the monitoring period while those for chronic exposure were calculated using the annual time-weighted average (ATWA) of estimated weekly ambient concentrations of MeBr at each receptor (Table 8.4). The CDPR uses an MOE of at least 100 to protect the health of the general public from acute exposure to MeBr (Lim, 1992). The U.S. EPA regards an MOE of 100 adequate to protect the health of the general population from chronic exposure to MeBr (IRIS, 1992).

The MOEs for acute exposure were about 10,000 and for chronic exposure were >100 for all monitoring stations. However, with the higher emission rate (Yates et al., 1996), the ISCST3-predicted concentrations are generally twice those measured and the CALPUFF predicted concentrations are 1.6 times the measured. Using these higher predicted concentrations, the MOE at Salinas lies close to 100. Thus, conditions of heavier than average usage in a given locale and exposures to infants and children and other sensitive population subgroups suggest that close attention should be given to ambient exposures in regions, such as the Salinas Valley, where the fumigant is widely used.

**TABLE 8.4**

Models for Translating Emissions into Ambient Concentrations in Comparison to the Risk-Based Reference Concentration

| Ambient Concentrations of MeBr in the Salinas Valley, CA ($\mu g/m^3$) | | | | |
|---|---|---|---|---|
| | **Salinas North** | **Salinas South** | **Arroyo Seco** | **King City** |
| Highest 24 hour Avg. Conc. Measured | 5.68 | 5.94 | 1.65 | 0.51 |
| Annual Time Wt. Avg. Conc. (WAC) | 2.48 | 2.51 | 0.76 | 0.29 |
| Worst Case (ISCST3)=2×WAC | 4.96 | 5.02 | 1.52 | 0.58 |
| Worst Case (CALPUFF)=1.66×WAC | 4.12 | 4.17 | 1.26 | 0.48 |
| NOAEL=81,600 kg/$m^3$ | | | LOAEL (HEC)=480 $\mu g/m^3$ | |

*Source:* Reprinted from Honaganahalli, P.S., and Seiber, J.N. (2000). Measured and predicted airshed concentrations of methyl bromide in an agricultural valley and applications to exposure assessment. *Atmos. Environ.* 34, 3511–3523. Copyright (2000), with permission from Elsevier.

*Note:* These are worst case ambient air concentrations for comparison with NOAEL (Lim, 1992) and lowest-observed-adverse-effect level (LOAEL) values (IRIS, 1992) from CalEPA Office of Environmental Health Hazard Assessment (OEHHA). Choosing the more conservative models provided an additional margin of safety.

Section 8.3.2 was adapted and partially (Reprinted from *Atmospheric Environment* 34, Honaganahalli, P.S., and Seiber, J.N., Measured and predicted airshed concentrations of methyl bromide in an agricultural valley and applications to exposure assessment, Pages 3511–3523, Copyright (2000), with permission from Elsevier.)

## 8.4 Modeling Emissions

Table 8.5 contrasts emission rates measured immediately after application for five different fumigants with different vapor pressures (VPs) and soil mobilities: chloropicrin, 1,3-D, MITC, methyl iodide (MeI), and MeBr with different application methods and depths (Woodrow et al., 2011). Emission rates of fumigants depend on their VP and soil diffusion rates ($S_w$); however, the same fumigant emits at different rates based on application method (see MITC drip, shank, surface chemical application), and the depth (see MeI, injected at three depths).

Numeric techniques that correlate pesticide physicochemical properties with empirically determined emissions such as these values can be used to estimate emission rates for different fumigants and different conditions of

**TABLE 8.5**

Approximate Volatilization Rates of Common Fumigants as Measured in the Field and Laboratory

| Fumigant | VP (Pa) | $S_w$ (mg/L) | Application Method | Application Depth (cm) | Measured Emission Rate ($\mu g/m^2/s$) |
|---|---|---|---|---|---|
| Chloropicrin | 3,173 | 2,270 | Shank | 26.6 | 98.4 |
| 1,3-D | 3,866 | 2,250 | Drip | 15 | 12.8 |
| 1,3-D | 3,866 | 2,250 | Shank | 38.1 | 9.03 |
| MITC | 2,666 | 7,600 | Drip | 10 | 4.06 |
| MITC | 2,666 | 7,600 | Shank | 22.9 | 2.58 |
| MITC | 2,666 | 7,600 | Surface chem | N/A | 75 |
| MeBr | 216,645 | 13,200 | Shank | 30 | 81 |
| MeI | 53,061 | 14,200 | Shank | 25.4 | 96.3 |
| MeI | 53,061 | 14,200 | Shank | 15.2 | 210 |
| MeI | 53,061 | 14,200 | Shank | 30.5 | 111 |

*Source:*　Woodrow et al. (2011).

use (Woodrow et al., 2011). A new correlation equation for emission rate was developed (8.4) where dispersion using wind speed and direction were modeled separately (Woodrow et al., 2011). This equation holds for chemicals of similar application depth and application methods under the same or similar application sites. The emission rates [ER ($\mu g/m^2/s$)] for subsurface injections and surface chemigations for 15 fumigant applications were combined with the physicochemical properties of the fumigants [VP (Pa); water solubility, $S_w$ (mg/L); soil adsorption coefficient, $K_{oc}$ (mL/g)] and with application conditions [application rate, AR (kg/ha); depth of application, d (cm)]. Resulting in the regression:

$$Ln\ ER = 3.598 + 0.9400\ Ln\left[(VP \times AR)/(S_w \times K_{oc} \times d)\right] \qquad (8.4)$$

Predictions by the equation were compared with measured results, with generally satisfactory results (Table 8.6). Cumulative loss and estimated values for different fumigants were composited from several experiments conducted at different times. The same methodology was then applied to predicting MeBr plumes.

## 8.5 Mitigation

Rapid biodegradation of MeBr in the soil makes trapping it there much more beneficial than release to the atmosphere where it is an ozone depleter (Shorter et al., 1995). There are a number of ways to reduce emissions of and exposures

**TABLE 8.6**

Comparison of Estimated with Measured Fumigant Emissions from Treated Fallow Fields

| Fumigant | Measured Emission Rate ($\mu g/m^2/s$) | Cumulative Losses | Estimated Emission Rate ($\mu g/m^2/s$) |
|---|---|---|---|
| MeBr | 43–86 | 22%–32% | 39–88 |
| Chloropicrin | 98–113 | 37%–60% | 96–109 |
| 1,3-D | 7–13 | 12% | 8–14 |
| MITC (drip, shank) | 1–5 | 4% | 2–4 |
| MITC (surf. chem.) | 43–75 | NA | 43–73 |

*Source:* Using Equation 8.4 and data from Woodrow et al. (2011).

to MeBr and other fumigants. Common mitigation strategies include use of high-density polyethylene (HDPE) tarps or virtually impermeable film (VIF), soil amendment with sodium thiosulfate, intermittent water seals, and use of buffer zones. In nontarped fields, 89% volatilized within 5 days (Majewski et al., 1995). However, with a heavy plastic tarp applied post injection, 22% was lost within the first 5 days of the experiment and about 32% within 9 days including (with tarp removed after 8 days).

## 8.6 Replacement

Another option is to switch to an entirely different chemical such as MITC, which can be applied via drip irrigation or shank injection and can further be contained via intermittent water seals and buffer zones.

### 8.6.1 Methyl Isothiocyanate

MITC is a fumigant of increasing use in agriculture (see Table 8.1). MITC is a gas that can replace some uses of MeBr. It has the advantage that it can be applied as a solid salt which serves as a precursor to gaseous MITC—either sodium or potassium isothiocyanate (metam sodium or potassium). This also makes it easier to ship and sell.

MITC is relatively safe for humans. It is a respiratory irritant, so it has its own built in warning system—eyes water and swell upon moderate exposure. This is reversible and full recovery occurs in case of an overdose or poisoning. It can be added as the salt to a drip irrigation system generating the fumigant MITC in situ, eliminating exposure to toxic gas and reducing the overall dosage and loss to the atmosphere. It does not threaten stratospheric ozone.

Since MITC is a natural product, released at ppm levels from broccoli and other cole crops, organic farmers make use of it by finely chopping waste broccoli and then adding it to soil or water to achieve some nematode and pathogen control effects from liberated MITC. Species of mustard have been selected for cultivation for use as a nematocidal powder, earning the title of "biofumigant" from Marrone BioInnovations. MITC use in California, primarily as MITC or its precursor salt metam sodium or potassium, has decreased from 8.2 million pounds in 2009 to 3.1 million pounds 2016 (Table 8.1). It is used primarily for treating strawberries, hops, and a number of other fruits and vegetables.

There have been sporadic reports of excessive exposure to people living around fields treated with MITC or a precursor. In one situation, a number of people near Bakersfield were exposed to higher-than-normal levels of MITC, apparently caused by a change in weather after application. Everyone recovered, but fines were levied for not adhering to weather restrictions while using MITC. As with all pesticides and all fumigants, the CDPR provides oversight in case of spills or exposures to people from use of pesticides in California.

The application technique can help minimize the risk of exposure. Woodrow and Seiber reported on MITC levels in air above fallow strawberry fields treated via direct fumigation in Orange County, California (Woodrow et al., 2008). Air levels were measured for several days after treatment. Levels were higher when the MITC was applied via shank injection vs when it was added to driplines.

Vince Hebert, at Washington State University, compared the use of shank injection of MITC with surface chemigation, a common practice among potato growers in Washington (Littke et al., 2013). The result was that shank injection led to lower emissions of the fumigant to air. It was recommended that potato growers use shank injection in addition to buffer zones. Emissions were even lower when only the beds were treated instead of the entire field. The highest concentrations observed were well below threshold limit values recommended by Cal EPA and Cal Occupational Safety and Health Administration (OSHA). Minimal noticeable levels of MITC were observed in air.

Unfortunately, the risks associated with MITC do not stop there. MITC is converted by oxidation to methyl isocyanate (MIC) in the air during and after release (Lu et al., 2014) (Figure 8.11). This is an issue for concern since MIC is more toxic than MITC. Its persistence is measured in half-lives calculated for MITC and MIC (Table 8.7).

MIC was the toxicant released during an industrial accident in Bhopal, India, in 1984, resulting in death or extreme illness to thousands of villagers in the downwind direction of the chemical plant where the accident occurred (Varma and Varma, 2005). Although this was not related to agricultural applications of MITC or precursor metam salts, the event reinforced the need for rigorous safety standards to protect agricultural workers and rural residents

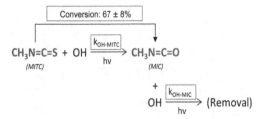

**FIGURE 8.11**
OH radical oxidation of MITC. (Data from Lu et al. 2014).

**TABLE 8.7**

Estimated Half-Lives of MITC and MIC (Hours)

| Test Compounds | Half-Lives (h)[a] |
| --- | --- |
| MITC | 15.6 |
| MIC | 66.5 |

*Source:* Reprinted (adapted) with permission from Lu, Z., Hebert, V.R., and Miller, G.C. (2014). Gas-phase reaction of methyl isothiocyanate and methyl isocyanate with hydroxyl radicals under static relative rate conditions. *J. Agric. Food Chem.* 62, 1792–1795. Copyright (2014) American Chemical Society.

[a] Half-lives calculated by using the average OH radical rate constants determined and an average OH radical concentration of $8.0 \times 10^5$ molecules/cm$^3$.

who might be exposed by accident to higher-than-expected levels of MIC/MITC. Bhopal was inundated by high levels (30–40 tons within 2 hours) of MIC from a ruptured tank. It is doubtful that agricultural workers would ever be exposed to such levels.

A train car derailment near Mt. Shasta and the California-Oregon Border in 1991 resulted in spill of MITC (as the precursor metam sodium) to a 50-mile stretch of the upper Sacramento River, northeast of Lake Shasta (Figure 8.12). There were no injuries, but many fish died when an estimated 19,000 gallons of metam sodium entered the river at the derailment site (Casey et al., 1993). Other fish recovered within weeks of the accident, apparently by swimming upstream and seeking refuge in creeks and ditches outside the main river channel. This incident shows reversibility of acute effects of MITC in animals and the ecosystem, but it still took 20 years for the rainbow trout population to recover.

## 8.6.2 Other Fumigants

MeI was proposed as a replacement for MeBr. They are chemically similar and act similarly as fumigants. MeI is not a stratospheric ozone depleter, but it is a potent neurotoxin, and thus has not been recommended by County

**FIGURE 8.12**
View of train derailment leading to the metam sodium contamination of the Sacramento River.
(© California Department of Fish and Game.)

Agricultural Commissioners in California. Hence, it is not used in California. It is registered in only a few other states and nations.

Telone or 1,3-D is another potential replacement for MeBr (Figure 8.1). Telone use has increased substantially in California as a result of threatened curtailment and/or banning of MeBr, as has a mix of Telone and chloropicrin, still another soil fumigant. The active ingredient in this mixture is 1,3-D with chloropicrin serving as the warning agent. Telone is more toxic to people than MeBr and some of the other replacements (Table 8.2). DPR, along with the California Air Resources Board (CARB), operates and maintains ambient monitoring stations for fumigants located near Chualar, California, in the Salinas Valley near where the application of 1,3-D to strawberries takes place each year. In the past few years, there have been occasions when the average air level of 1,3-D has exceeded expected levels at those reporting stations. DPR, CARB, and other officials are investigating the reason for these unacceptable breaches. DPR and CARB results are reported at the California Pesticide Registration Evaluation Committee meetings as occasional updates.

Nonresidual fumigants such as ethylene oxide and propylene oxide are often used to treat perishable commodities prior to shipment abroad. Fruits, nuts, hops, and other crops are fumigated to kill termites, ants, fungi, and other vermin in order to avoid losses in transit. This is often done in shipping containers near the point of departure at ports and warehouse terminals. Without proof of treatment, these containers may be turned away at customs at the end of their journeys and not marketed. Even nonfood commodities such as wood and lumber may need to be treated to kill termites and other hitchhiking pests in furniture and clothing. How long to treat them, and at what quantity or pressure are variables that are worked out in advance. USDA-APHIS and ARS are the lead agencies for determining the conditions for these quarantine treatments. See the homepage for sites such as the ARS San Joaquin Valley Agricultural Sciences Center in Parlier, California, and the West Side Research and Extension Center in Five Points, California.

Biofumigants are increasingly being tested as replacements for synthetics like MeBr (Figure 8.13). The mustard plants of Brassica are a source of biofumigant isothiocyanates, and Marrone BioInnovations (Davis, CA) is in the process of commercializing nematicide formulations based on biofumigants. Another biofumigant, or green fumigant, is ethyl formate, which acts by acylating pest DNA but with minimal side effects that might endanger humans and other mammals. Other natural sources of nematicides include α-terthienyl, which is produced in marigolds. This chemical is not commercialized but is well known among organic gardeners. The Dow Chemical Company considered marketing α-terthienyl or its derivatives but was unable to find an economical synthetic process starting with thiophene—a low-cost chemical by-product of refining petroleum. Tetrachlorothiophene is a registered nematicide but no longer used, also because it is not considered commercially viable.

Two synthetics also have nematode control uses. One is Nemacur®, an organophosphate marketed by Bayer Crop Science. Another is avermectin, marketed by Merck for agricultural use (see Chapter 12). A different formulation of avermectin is used as a veterinary medicine to treat worms and flukes in cattle, reindeer, and sheep. Avermectin is a biopesticide as defined

**Alternative Pest Control Agents:**
**3rd Generation**

*Biopesticides based on naturally occurring chemicals

*Semiochemicals: Insect communication with chemicals

*JH Analogues: Juvenile Hormones from insects used as
    "Lead compounds"

*Genetically Modified Crops (GMOs)
    -Insecticidal crops (e.g., Bt-crops)
    -Herbicide-resistant crops (e.g., Roundup Ready crops)

**FIGURE 8.13**
Alternatives to fumigants are considered third-generation pesticides.

**Bob Krieger's Legacy**

*Personal Chemical Exposure Program – human exposure assessment in risk perception and risk management

*His tenacity in addressing chemical safety issues

*His former students, many of whom went on to assume major leadership positions in industry, government, and academia

*His memory as a person who was soft-spoken, kindhearted, and generous with his time

**FIGURE 8.14**
This chapter dedicated to Bob Krieger, a tireless advocate for safety in the use of fumigants and other pesticides. He inspired much of the work described here.

by the EPA, since it is made by fermentation of avermitilis species and occurs naturally. It is expensive for agricultural use, and it has the potential to leave residues if used on food crops, but it can be an alternative to fumigation and can be used in planted strawberry fields in addition to fallow fields preplant. Development of new candidate nematicides is accelerating, due to the potential market demand with strawberries, hops, and other high value crops.

This chapter is dedicated to our friend and colleague, Bob Krieger (Figure 8.14).

## References

Anderson, T.A., Rice, P.J., Cink, J.H., and Coats, J.R. (1996). Fate of methyl bromide in fumigated soils. In *Fumigants*, (pp. 42–52). Washington, D.C.: American Chemical Society.

Babich, H., Davis, D.L., and Stotzky, G. (1981). Dibromochloropropane (DBCP): A review. *Sci. Total Environ.* 17, 207–221.

California Department of Pesticide Regulation (CDPR). (2018). The top 100 pesticides by pounds in total statewide pesticide use in 2018 https://www.cdpr.ca.gov/docs/pur/pur18rep/top_100_ais_lbs_2018.htm (Accessed 2/3/21).

Casey, A., Fenster, L., and Neutra, R. (1993). *An Investigation of Spontaneous Abortions Following a Metam Sodium Spill into the Sacramento River.* Sacramento, CA: California Department of Health Services, Environmental Health Investigations Branch.

Honaganahalli, P.S., and Seiber, J.N. (1996). Health and environmental concerns over the use of fumigants in agriculture: The case of methyl bromide. In *Fumigants*, J.N. Seiber, J.E. Woodrow, M.V. Yates, J.A. Knuteson, N.L. Wolfe, and S.R. Yates, (Eds.), (pp. 1–13). Washington, DC: American Chemical Society.

Honaganahalli, P.S., and Seiber, J.N. (2000). Measured and predicted airshed concentrations of methyl bromide in an agricultural valley and applications to exposure assessment. *Atmos. Environ.* 34, 3511–3523.

IRIS. (1987). Carbon disulfide (CASRN 75-15-0) | IRIS | US EPA. https://iris.epa.gov/static/pdfs/0217_summary.pdf (Accessed 2/12/2021).

IRIS. (1992). Integrated risk information system. Bromomethane, CASRN 74-83-9. https://iris.epa.gov/static/pdfs/0015_summary.pdf. (Accessed 2/12/2021).

Lee, S., McLaughlin, R., Harnly, M., Gunier, R., and Kreutzer, R. (2002). Community exposures to airborne agricultural pesticides in California: Ranking of inhalation risks. *Environ. Health Perspect.* 110, 1175–1184.

Lim, L.O. (1992). *Methyl bromide preliminary risk characterization. Memorandum to Nelson*, L., dated 11 February 1992. Sacramento, CA: Medical Toxicology, Branch, California Department of Pesticide Regulation.

Littke, M.H., LePage, J., Sullivan, D.A., and Hebert, V.R. (2013). Comparison of field methyl isothiocyanate flux following Pacific Northwest surface-applied and ground-incorporated fumigation practices: Comparison of field methyl isothiocyanate flux following different fumigation practices. *Pest Manag. Sci.* 69, 620–626.

Lu, Z., Hebert, V.R., and Miller, G.C. (2014). Gas-phase reaction of methyl isothiocyanate and methyl isocyanate with hydroxyl radicals under static relative rate conditions. *J. Agric. Food Chem.* 62, 1792–1795.

Majewski, M.S., Glotfelty, D.E., and U, K.T.P (1990). A field comparison of several methods for measuring pesticide evaporation rates from soil. *Environ. Sci. Technol.* 24, 1490–1497.

Majewski, M.S., McChesney, M.M., Woodrow, J.E., Prueger, J.H., and Seiber, J.N. (1995). Aerodynamic measurements of methyl bromide volatilization from tarped and nontarped fields. *J. Environ. Qual.* 24, 742–752.

NASA. (1997) *Chemical Kinetics and Photochemical Data for Use in Stratospheric Modeling.* Evaluation No. 10: NASA panel for data evaluation. Pasadena, CA: California Institute of Toxicology, Jet Propulsion Laboratory, National Aeronautics and Space Administration.

Nelson, L. (1992). *Methyl Bromide Preliminary Risk Characterization.* Memorandum to Jim Wells dated 11 February 1992. Sacramento, CA: Medical Toxicology, Branch, California Department of Pesticide Regulation.

Phillips, T.W., Thoms, E.M., DeMark, J., and Walse, S. (2012). *Stored Product Protection.*

Poulet, G., Pirre, M., Maguin, F., Ramaroson, R. and LeBras, G. (1992). The role of BrO+ HO2 reaction in the stratospheric chemistry of Bromine. *Geophys. Res. Lett.* 19, 2305–2308..

Seiber, J.N., Woodrow, J.E., Honaganahalli, P.S., LeNoir, J.S., and Dowling, K.C. (1996). Flux, dispersion characteristics, and sinks for airborne methyl bromide downwind of a treated agricultural field. In *Fumigants*, (pp. 154–177). Washington, D.C.: American Chemical Society.

Shorter, J.H., Kolb, C.E., Crill, P.M., Kerwin, R., Talbot, R., Hines, M., and Harriss, R. (1995). Rapid degradation of atmospheric methyl bromide in soils. *Nature* 377, 717–719.

United States Environmental Protection Agency. (1992). Ethylene Dibromide (Dibromoethane). https://www.epa.gov/sites/production/files/2016-09/documents/ethylene-dibromide.pdf (Accessed 12/07/2020).

Varma, R., and Varma, D. (2005). The Bhopal Disaster of 1984. *Bull. Sci. Technol. Soc.* 25, 37–45.

Walse, S. Fumigation: USDA ARS. https://www.ars.usda.gov/pacific-west-area/parlier/sjvasc/cpq/people/spencer-walse/fumigation/ (Accessed 2/12/2021).

Woodrow, J.E., Seiber, J.N., and Baker, L.W. (1997). Correlation techniques for estimating pesticide volatilization flux and downwind concentrations. *Environ. Sci. Technol.* 21, 523–529.

Woodrow, J.E., Seiber, J.N., LeNoir, J.S., and Krieger, R.I. (2008). Determination of methyl isothiocyanate in air downwind of fields treated with metam-sodium by subsurface drip irrigation. *J. Agric. Food Chem.* 56, 7373–7378.

Woodrow, J.E., Seiber, J.N., and Miller, G.C. (2011). Correlation to estimate emission rates for soil-applied fumigants. *J. Agric. Food Chem.* 59, 939–943.

Woodrow, J.E., LePage, J.T., Miller, G.C., and Hebert, V.R. (2014). Determination of MIC in outdoor residential air near metam-sodium soil fumigations. *J. Agri. Food Chem.* 62, 8921–8927.

Yates, S.R. (2006). Measuring herbicide volatilization from bare soil. *Environ. Sci. Technol.* 40, 3223–3228.

Yates, S.R., Gan, J., Ernst, F.F., Wang, D., and Yates, M.V. (1996). Emissions of methyl bromide from agricultural fields: Rate estimates and methods of reduction. In *Fumigants*, J.N. Seiber, J.E. Woodrow, M.V. Yates, J.A. Knuteson, N.L. Wolfe, and S.R. Yates, (Eds.), (pp. 116–134). Washington, DC: American Chemical Society.

Yung, Y.L., Pinto, J.P., Watson, R.T. and Sander, S.P. (1980). Atmospheric bromine and ozone perturbations in the lower stratosphere. *J. Atmos. Sci.* 37, 339.

# 9

# Trifluoroacetic Acid from CFC Replacements: An Atmospheric Toxicant Becomes a Terrestrial Problem

## 9.1 Introduction

Our research on methyl bromide, a stratospheric ozone depleter, attracted the attention of the Alternative Fluorocarbons Environmental Acceptability Study (AFEAS), a consortium of chemical companies (Dupont, Imperial Chemical Industries (ICI), and others) with a common interest of developing alternatives to the chlorofluorocarbon (CFC) refrigerants and propellants. The CFCs depleted the stratospheric ozone layer which protects earth from higher energy ultra-violet light. They were also greenhouse gases that keep the atmosphere from radiating excess heat from the atmosphere to the stratosphere, and thus causing a rise in the earth's temperature leading to global warming. The CFCs were banned under an international agreement known as the Montreal Protocol. The primary greenhouse gas is $CO_2$, and nations are struggling to reduce $CO_2$ emissions with limited success. CFCs were GHGs that could be controlled, so most nations signed on to the Montreal Protocol as a first step in controlling global warming.

The AFEAS consortium sought to replace CFCs with other fluorocarbons that could be used in air conditioners, refrigerators, etc., but were not themselves "greenhouse gases." After much research and development, the replacement chemicals advanced by AFEAS were the hydrochlorofluorocarbons (HCFCs), such as $CF_3CHFCl$ (Figure 9.1). The presence of the C–H bond in these HCFCs allowed the HCFCs to degrade in the troposphere, and not reach the stratosphere, and thus not deplete naturally occurring ozone in the stratosphere. The earth would not warm irreversibly and life could continue.

But the HCFCs had a drawback: when they degrade in the atmosphere, the major product is trifluoroacetic acid (TFA). Although not very toxic, TFA is unusually stable and water soluble; TFA can build up in earth's waters that undergo evaporation to potentially harmful levels. AFEAS funded our group at University of Nevada, Reno, and University of California, Davis, to

DOI: 10.1201/9781003217602-9

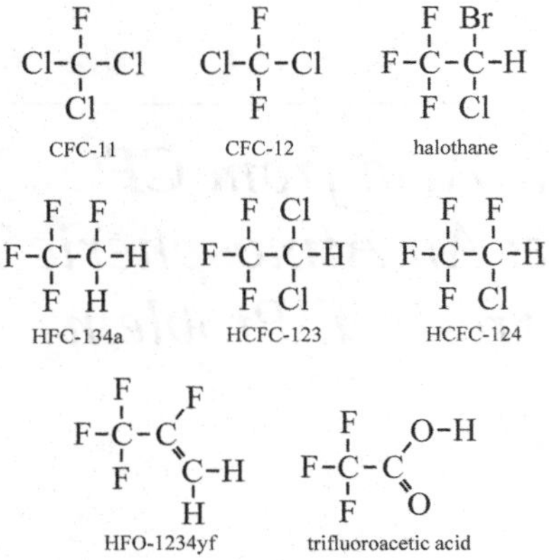

**FIGURE 9.1**
Structures of CFCs and HFCs. The first row are CFCs that have been phased out although halothane still has limited usage in developing countries. The second row are some of the first-generation CFC replacements, namely HFCs and HCFCs. The last row is a second-generation HFC, HFO-1234yf, and TFA. All of the compounds shown except for CFC-11 and CFC-12 can form TFA.

investigate the lifetime and fate of TFA in the biosphere. Studies of Tom Cahill and others in our group, plus the relevant literature, are reported in Chapter 9.

Bottomline: we found the expected buildup in terminal waterbodies like Mono Lake, and to a lesser extent in the oceans, but the TFA buildup did not achieve levels high enough to harm the biosphere or its flora and fauna.

## 9.2  TFA as an Environmental Concern

The concern over TFA originally emerged during the replacement of chlorofluorocarbons (CFCs) with hydrofluorocarbons (HFCs) for use as refrigerants, foam blowing agents, aerosol propellants, and some fire suppression systems (Figure 9.1). Prior to the 1980s, the most common refrigerants were the CFCs. These compounds were ideal for this application because they were inert gases that would not break down in the refrigeration unit. They were also nontoxic and nonflammable, unlike some of the other early refrigerants such as ammonia, chloromethane, and sulfur dioxide. The stability of the CFCs, while a benefit in their application, represented a problem in the atmosphere since they were minimally degraded in the troposphere. The CFCs could then slowly cross the tropopause and enter the stratosphere where they were exposed to more energetic ultraviolet energy with wavelengths less than 227 nm (Molina and Rowland, 1974). The carbon–chlorine bond could then photolyze which resulted in the release of a chlorine radical. The chlorine radical would then catalytically destroy the ozone by the following reactions (Molina and Rowland, 1974):

$$Cl\bullet + O_3 \rightarrow ClO\bullet + O_2 \tag{9.1}$$

$$ClO\bullet + O \rightarrow Cl\bullet + O_2 \tag{9.2}$$

Therefore, the net reaction is:

$$O_3 + O \rightarrow O_2 + O_2 \tag{9.3}$$

In these reactions, the free chlorine is a catalyst that is continuously regenerated and a single chlorine atom can destroy considerable amounts of ozone until it reaches a chain termination reaction, such as chlorine radical reacting with methane to create hydrochloric acid and a methyl radical. Bromine radicals from brominated compounds, such as halothane, were even more efficient in catalyzing ozone destruction. The loss of ozone in the stratosphere resulted in a thinning of the ozone layer and ozone holes over the Polar Regions and the Antarctic region in particular (Farman et al., 1985). If the thinning of the ozone layer was not reversed, then the absence of ozone would allow more energetic ultraviolet light to reach the earth's surface which could lead to detrimental effects such as increased skin cancer and global warming.

In 1987, the Montreal Protocols were established to rapidly phase out the production and use of CFCs in industrialized countries by the year 2000 and in developing countries by 2010, although illegal, new production of CFC-11 has been detected as recently as 2017 (Montzka et al., 2018; Rigby et al., 2019). The HCFCs were utilized as a short-term replacement that was ultimately scheduled to be phased out as well. The main CFC replacements were the HFCs, such as HFC-134a. These compounds differed from the CFCs in two important respects. The first was that the HFCs lacked any chlorine or bromine atoms that could catalyze ozone destruction. The second feature of these compounds was that they contained one or more hydrogen atoms that could react with hydroxyl radicals in the atmosphere (Wallington et al., 1994). Therefore, the HFC chemicals could degrade in the troposphere and may not make it into the stratosphere to the same extent as the CFCs. Some of the HFCs, such as HFC-134a, were also potent greenhouse gases, so they are being phased out in favor of other HFCs that have a lower global warming potentials (GWPs), such as HFO-1234yf.

The HFCs were designed to be less stable than the CFCs that came before them, so some of them could degrade in the troposphere before reaching the stratosphere. However, the degradation rate of the first generation of HFCs and HCFCs was still very slow as exemplified by the atmospheric lifetimes ($\tau$) of HFC-134a, HCFC-123, and HCFC-124: 14.4, 1.4, and 6.2 years, respectively (Kotamarthi et al., 1998). For most HFCs, the first reaction is a hydrogen abstraction by a hydroxyl radical (Franklin, 1993; Kotamarthi et al., 1998) (Figure 9.2).

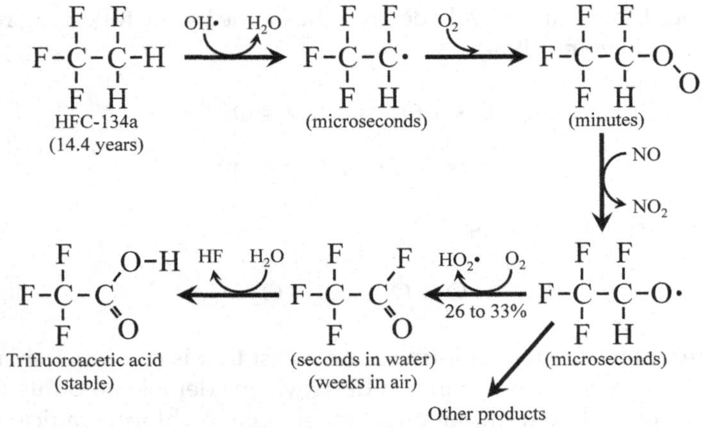

$$
\underset{\substack{\text{HFC-134a}\\(14.4\text{ years})}}{F\text{-}\underset{F}{\overset{F}{C}}\text{-}\underset{H}{\overset{F}{C}}\text{-}H} \xrightarrow[H_2O]{OH\cdot} \underset{(\text{microseconds})}{F\text{-}\underset{F}{\overset{F}{C}}\text{-}\underset{H}{\overset{F}{C}}\cdot} \xrightarrow{O_2} \underset{(\text{minutes})}{F\text{-}\underset{F}{\overset{F}{C}}\text{-}\underset{H}{\overset{F}{C}}\text{-}O\text{-}O\cdot}
$$

NO → NO₂

$$
\underset{\substack{\text{Trifluoroacetic acid}\\(\text{stable})}}{F\text{-}\underset{F}{\overset{F}{C}}\text{-}\overset{O\text{-}H}{\underset{O}{C}}} \xleftarrow[H_2O]{HF} \underset{\substack{(\text{seconds in water})\\(\text{weeks in air})}}{F\text{-}\underset{F}{\overset{F}{C}}\text{-}\overset{F}{\underset{O}{C}}} \xleftarrow[26\text{ to }33\%]{HO_2\cdot\ \ O_2} \underset{(\text{microseconds})}{F\text{-}\underset{F}{\overset{F}{C}}\text{-}\underset{H}{\overset{F}{C}}\text{-}O\cdot}
$$

Other products

**FIGURE 9.2**
Formation of TFA from HFC-134a. Adapted from Franklin (1993) and Kotamarthi et al. (1998). The approximate atmospheric lifetimes of the chemicals are given under each structure.

This is the slowest reaction in the degradation of these compounds; the rest of the reactions occur on the microsecond to minute time scale. One of the major products from the degradation of these compounds is TFA (Wallington et al., 1994). Since the atmospheric lifetime of these HFCs/HCFCs was measured in years, the formation of TFA was expected to be distributed around the globe. The more recent HFCs, such as HFO-1234yf, are more reactive and can degrade relatively quickly with a lifetime of approximately 11 days (Nielsen et al., 2007), but it still results in the formation of TFA. Therefore, the degradation of HFO-1234yf will result in TFA formation on a regional scale rather than a global scale (Henne et al., 2012; Luecken et al., 2010).

The TFA formed from the degradation of HFCs was expected to be exceptionally stable and effectively immune to further oxidation under typical environmental conditions. Based on its physicochemical properties (discussed in detail later), it was expected to wash out of the atmosphere and enter surface waters as a nonvolatile trifluoroacetate ion. Given the stability and complete lack of volatility of TFA, it was feared that it could accumulate in waterbodies lacking outflows until toxic concentrations could be achieved (Russell et al., 2012; Schwarzbach, 1995; Tromp et al., 1995). This initiated the research of TFA as an environmental pollutant.

## 9.3 Sources of TFA

The initial source of TFA was suspected to be the atmospheric degradation of the HFC compounds. However, it has since been realized that there were many additional sources of TFA. Most chemicals that have a trifluoromethyl functional group ($-CF_3$) attached to a carbon are now suspected of potentially forming TFA. Not all chemicals with a trifluoromethyl group will form TFA since there are some other products that can be formed, such as trifluoromethanol that can eventually be converted to $CO_2$ and HF (Franklin, 1993). The degree to which a $-CF_3$ containing molecules will form TFA is dependent on the particular chemical, but most of them have the potential to form TFA to some degree. Some of these additional chemicals that can form TFA have the potential to create localized increases in TFA concentrations.

The first, and probably largest source of TFA, is the atmospheric oxidation of HFC compounds as discussed above. These HFCs and HCFCs have many applications. For example, HFC-134a is the main replacement for CFC-12 and it is used extensively in small refrigeration units in cars and domestic situations. Since this compound is not flammable, it is used as a propellant in aerosol cans and "canned air" for removing dust from electronics. HFC-134a has a high GWP of 1,200 compared to $CO_2$ which has a GWP of 1.0 by definition. Due to this high GWP, HFC-134a is currently being phased out in favor of compounds like HFO-1234yf that have considerably lower GWP values due to their short half-life in the atmosphere. Some HCFCs, such as HCFC-123 and HCFC-124, have likewise been used as transitional substitutes for the CFCs. These compounds, like HFC-134a, also form TFA when they degrade in the atmosphere (Kotamarthi et al., 1998). These compounds also have high GWP, so they are scheduled to be phased out of new products in 2020 in developed countries and 2030 in developing countries although they can be used to service existing equipment for an additional ten years. It is worth noting that not all transitional HFCs and HCFCs form TFA. For example, HCFC-22 ($CHClF_2$) was the most used refrigerant in the 2010s (Booten et al., 2020), but it does not form TFA simply because it is a single carbon with two fluorines, one chlorine and a hydrogen on it. HCFC-141b ($CH_3CCl_2F$) and HCFC-142b ($CH_3CClF_2$) are other examples. HFCs and HCFCs that have the potential to form TFA must have a trifluoro group in the molecule.

The atmospheric concentrations of these compounds, and HFC-134a in particular, have increased as their usage has increased (Figure 9.3). The global consumption of HFC-134a was estimated to be between 260,000 and 310,000 t/yr, which makes it the most abundant currently used refrigerant second only to HCFC-22 ($CHClF_2$) (Booten et al., 2020). It is expected that TFA generation in the atmosphere will increase alongside the increased atmospheric concentrations of HFC-134a. While HFC-134a is still a major refrigerant, its phaseout has already begun and its most probable replacement is HFO-1234yf (Figure 9.1), which has a low GWP due to its very short atmospheric half-life. By 2021, all new vehicles in the United States will have HFO-1234yf as the refrigerant instead of HFC-134a. HFO-1234yf has solved both the ozone-depleting problem and the global warming problem, but it still forms TFA in the process (Henne et al., 2012; Luecken et al., 2010).

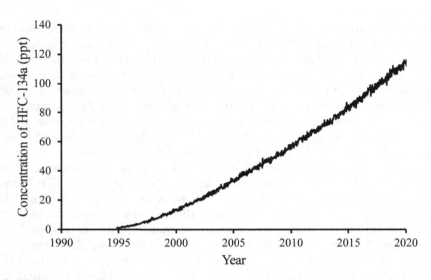

**FIGURE 9.3**
Atmospheric concentrations (ppt) of HFC-134a at Mauna Loa Observatory, Hawaii. (Data from the NOAA/ESRL Global Monitoring Laboratory, Boulder, CO, USA.)

Inhaled anesthetics represent another source of halocarbons that can form TFA when they are released to the environment. The anesthetic halothane ( 2-bromo-2-chloro-1,1,1-trifluoroethane) is still used in developing countries although it has been mostly replaced by other chemicals. In 2014, it was estimated that 250 t/yr of halothane was being used (Vollmer et al., 2015) even though its ozone-depleting potential (ODP) is 1.56, which is greater than that of CFC-11. Its high ODP is largely due to the presence of a bromine atom that is even more efficient at catalyzing ozone destruction than chlorine atoms. Most developed countries have switched over to fluorinated compounds such as isoflurane, sevoflurane, and desflurane (Vollmer et al., 2015). In 2014, the emissions of isoflurane, sevoflurane, and desflurane were estimated to be 880, 1,200, and 960 t/yr, respectively. These compounds combined represent approximately 1% of the abundance of HFC-134a consumed each year, so they are minor compared to the HCF refrigerants. Unfortunately, all of these anesthetics, along with halothane, have the potential to form TFA (Andersen et al., 2012; Hankins and Kharasch, 1997; Wallington et al., 2002). In contrast to refrigerants that tend to leak out of products slowly, the anesthetics are completely released during each usage. No significant efforts to recapture these compounds after use have been made. There are alternatives to these inhaled anesthetics, such as injected anesthetics, xenon (Xe) and nitrous oxide (N$_2$O), but they have their own limitations.

Another identified source of TFA is the thermolysis of fluoropolymers such as polytetrafluoroethylene (PTFE) (Ellis et al., 2001b). These plastics are specifically used in high temperature situations where the stability of the plastic is essential. The mechanism appears to be the breakdown of the fluoropolymer into difluorocarbene units ($\bullet CF_2 \bullet$) that can then condense to form hexafluoropropene ($CF_3-CF=CF_2$). The double bond in this molecule can then be epoxidized, which then rearranges to form trifluoroacetyl fluoride ($CF_3-C(O)-F$). This acid halide then hydrolyzes in water to form TFA (Ellis et al., 2001b). In addition to TFA, some longer chain perfluorinated carboxylic acid species may also be generated. Additionally, fluoropolymers destroyed in incinerators might create the volatile precursor gases that could transform into TFA. This could explain the abnormally high concentrations of TFA in urban areas that could not be attributed to HFC degradation when HFCs were starting to be utilized.

The degradation of long-chain perfluorinated surfactants and fluorotelomer alcohols (Ellis et al., 2004) can also give rise to TFA. Compounds such as perfluorooctanesulfonate (PFOS) and perfluorooctanoic acid (PFOA) are persistent, toxic, and bioaccumulate in the environment. Considerable research has been dedicated to devise mechanisms to degrade these compounds in wastewater treatment plants, so they are not released to the environment. Most of the strategies for degrading these chemicals involves the stepwise removal of carbons starting at the acid end of the molecule. This creates a series of shorter chain perfluorinated acids as intermediates which can result in the production of some TFA (Hori et al., 2005; Singh et al., 2019) although many studies do not measure TFA so the exact contribution of the perfluorinated acids in making TFA is unclear.

An unexpected source of TFA was the degradation of certain pesticides containing a trifluoro functional group. 3-Trifluoromethyl-4-nitrophenol (TFM), used to control lamprey in the Great Lakes of the United States and Canada, has been shown to degrade into TFA (Ellis and Mabury, 2000). This compound is used at a rate of 50 metric tons per year in the Great Lakes. Trifluralin, a preemergent herbicide, has also been suggested as a potential source of TFA, but it has not been definitively demonstrated (Jordan and Frank, 1999). The pesticides flurtamone (herbicide), fluopyram (fungicide), tembotrione (herbicide), and flufenacet (herbicide) have been shown to generate TFA during ozonation treatment of wastewater (Scheurer et al., 2017).

Pharmaceuticals represent another potential source of TFA. In this case, the pharmaceuticals are used and end up in the wastewater where they can be degraded, either naturally or by wastewater treatment programs, to form TFA. Fluoxetine (trade name Prozac) and sitagliptin (trade name Januvia) are known to form TFA during the ozonation of wastewater (Scheurer et al., 2017).

The last, and somewhat debated, source of TFA is nature. Very early research showed TFA present in rivers and seawater in Europe at high concentrations early in the adoption of the HFC replacement transition, so they could not be explained by HFC oxidation (Frank et al., 1996). Further analysis of old well water showed that TFA was largely undetectable, giving rise to the assessment that "TFA seems to be predominantly if not exclusively of anthropogenic/industrial origin" (Jordan and Frank, 1999). This result was reinforced by Nielsen et al. (2001), who studied ancient fresh water, from both groundwater and Greenland ice cores, and found undetectable concentrations of TFA. In another study, ancient spring water also had no detectable TFA in it (<5 ng/L) (Berg et al., 2000). However, Von Sydow et al. (2000) measured Antarctic ice cores and detected relatively high concentrations of TFA in ice that predates the industrial age, so they concluded that there must be a natural source of TFA. Additionally, studies of ocean water columns showed consistent concentrations of TFA near 200 ng/L even in deep seawater greater than 60 years old (Frank et al., 2002). The authors conclude that the freshwater TFA is predominately anthropogenic in origin while the TFA in seawater arises from some natural source that concentrates in the ocean due to TFA's great stability. Additional ocean research has confirmed relatively high TFA concentrations in deep ocean water that predated the industrial age (Scott et al., 2005). This research also sampled water over deep-sea geothermal vents that showed elevated TFA, which suggests that deep-sea geothermal vents are a source of TFA. Some fluorinated organics have been detected in fluorite and volcanic rocks (Harnisch et al., 2000), but the fluorinated compounds identified so far have been limited to single carbon or single sulfur molecules that lack the ability to form TFA.

## 9.4 Chemistry of TFA

As with all chemicals, the physicochemical properties of TFA dictate its fate and transport in organisms and the environment.

### 9.4.1 Acidity

The first and most important characteristic of TFA is its strong acidity that causes it to be an ion under environmental conditions. This strong acidity, as quantified by the low $pK_a$ of TFA (Table 9.1), is due to the electron-withdrawing fluorines in the molecule. When comparing acetic acid to the TFA series of compounds, it is clear that each fluorine addition lowers the $pK_a$ and therefore makes a stronger acid.

The low $pK_a$ for TFA means that it will readily ionize in water under almost any environmental condition to become the trifluoroacetate ion, which has several implications for its environmental fate and transport. If TFA is generated in the atmosphere, it will partition to water (rain, fog), ionize, and be washed out of the atmosphere. Bowden et al. (1996) measured the Henry's law constant for TFA ($K_H$=8,950±100 mol/kg/atm at 298 K) and determined that the ionization of TFA was the main property that causes TFA to partition into atmospheric water where it could be

**TABLE 9.1**

$pK_a$ of the Fluorinated Acetic Acid Compounds

| Chemical | $pK_a$ |
| --- | --- |
| Acetic acid | 4.80[a], 4.76[b] |
| Fluoroacetic acid | 2.66[a], 2.59[b] |
| Difluoroacetic acid | 1.24[a] |
| TFA | 0.23[a], 0.52[b], 0.1 to 1.0 (modeled)[c] 0.19[d] |

[a] Data summarized in Richard and Hunter (1996).
[b] Data summarized in Lide (2003).
[c] Reference Namazian et al. (2008).
[d] Calculated from dissociation constant at 25°C presented in Milne and Parker (1981).

subsequently removed by wet deposition. Partitioning into aerosols was a more complex issue that was highly dependent on aerosol pH, where TFA would not significantly partition into highly acidic aerosols (Bowden et al., 1996). More recent measurements of the Henry's law constant gave a similar value of $5{,}800 \pm 100$ mol/dm$^3$/atm at 298 K (Kutsuna and Hori, 2008), which confirmed the propensity of TFA to partition out of the gas phase into the liquid phases. However, field measurements that divided the atmospheric TFA into gas-phase TFA and aerosol-phase TFA have repeatedly shown higher concentrations in the gas phase (Guo et al., 2017; Martin et al., 2003; Wu et al., 2014). Guo et al. (2017) showed that adsorption of TFA to particles was energetically favored but appears to be limited by diffusion processes. The partitioning of TFA to aerosols is a mechanism by which dry deposition of TFA can occur and this deposition may be as important as wet deposition (Martin et al., 2003).

The strong propensity of TFA to be an ion has implications for its behavior in terrestrial and aquatic systems. In general, terrestrial and aquatic systems have more moderate pH conditions (pH 5–8), so TFA will effectively be present as an ion. At pH 7.0, there would be approximately 3 million molecules in the ionic form for each molecule in the acid form. As an ion, TFA is effectively nonvolatile; therefore, it does not revolatilize to the atmosphere. TFA's high water solubility differentiates it from the longer chain perfluorinated acids, such as PFOS and PFOA. Once TFA is dissolved in water, it will predominately stay in the aqueous phase and be moved by the water to the oceans. The one notable exception to this is TFA inputs into terminal waterbodies, which are lakes, playas, and wetlands, that lack an outflow. In these cases, the ultimate fate of the water in this scenario is to be lost by evaporation while the TFA ion is left behind. This could cause the concentrations of TFA to increase over time in these systems as rain, dry deposition, and surface water inputs continually add more TFA to the system that cannot be lost by volatilization or degradation (Tromp et al., 1995). In many ways, TFA behaves in a similar fashion as some small inorganic ions such as chloride and bromide; it is highly water soluble, stable, and does not form any significant precipitates or complexes.

The predominately ionic character also affects the fate of TFA within organisms. Since TFA is a small, unreactive, ionic molecule, it will preferentially stay in the aqueous phases in animals where it can be readily excreted by the kidneys. The easy excretion of TFA in urine prevents it from bioaccumulating in animals like PFOS or PFOA. In plants, TFA can be acquired by direct deposition on foliage or TFA can be taken up in water and transported to plant tissues by the xylem. When the water evaporates from a plant, or is otherwise consumed during photosynthesis, TFA is left behind in the plant tissues. This could lead to a steady increase in tissue concentrations over time in the leaves. The TFA might be lost from plants by leaching into rainfall and some plants can excrete water by guttation.

### 9.4.2 Stability

The second physicochemical property of TFA is that it is exceptionally stable under most environmental conditions. Since TFA is a highly oxidized molecule, it is expected to be resistant to degradation by oxidation. Additionally, the carbon–fluorine bond is one of the strongest single bonds found in organic molecules, which makes it more difficult to break. It also lacks any hydrogen atoms that are vulnerable to hydrogen abstraction. Therefore, degradation in the environment is expected to be very slow.

Oxidation reactions are a common means to degrade organic chemicals, but their usefulness in degrading the highly oxidized TFA was uncertain. In the atmosphere, the gas-phase protonated form of TFA can react with hydroxyl radicals (OH•). The atmospheric lifetime of TFA due to the reaction with OH• has been estimated to be about 230 days (Hurley et al., 2004), 68 days (Mogelberg et al., 1994), and 100 days (Carr et al., 1994). Gas-phase reaction with TFA was slower than that of other perfluorinated carboxylic acids (PFCAs) (Hurley et al., 2004), so TFA is even more stable than other longer chain PFCAs. However, the researchers noted that TFA would most likely be removed from the atmosphere by wet or dry deposition with a lifetime of roughly 10 days, so atmospheric oxidation of TFA is a relatively minor loss process.

TFA degradation in aqueous systems under environmental conditions is expected to be very slow. Hydroxyl radicals in aqueous solutions are not very reactive toward TFA; TFA was the least reactive of the haloacetic acids (Maruthamuthu et al., 1995). Additionally, degradation of TFA in natural surface waters is likely to be slow due to the presence of other dissolved organic matter that would more readily react with hydroxyl radicals. In microcosm studies, there was no observed TFA degradation over 49 days, so a half-life could not be calculated (Hanson et al., 2002). A longer, 1-year study also did not detect any TFA degradation in pond water (Ellis et al., 2001a).

Thermal decomposition of TFA in water is likewise slow at ambient temperatures. There was no observed degradation of TFA at 90°C over 100 days

(Lifongo et al., 2010). If the maximum possible reaction rate for TFA at 90°C in this study was extrapolated to 15°C, then the half-life of TFA at this temperature would be 40,000 years, which means that simple thermal degradation would not be a factor for TFA. Currently, there is extensive research on mechanisms to degrade PFCAs, including TFA, in wastewater treatment plants to prevent their release to the environment. However, these processes are engineering solutions that are not applicable to environmental conditions (for examples, see Hori et al., 2005; Qian et al., 2016; Singh et al., 2019).

The terrestrial environment also has mechanisms to degrade TFA. Pehkonen et al. (1995) reported that TFA could be photooxidized by iron oxyhydroxides to generate $CO_2$ and a $F_3C\bullet$ radical that can subsequently be mineralized. However, the reaction rate for TFA was slower than that for fluoroacetate or difluoroacetate using two iron oxyhydroxides, namely, ferrihydrite $\left(Fe^{+3}_5O_3(OH)_9\right)$ and maghemite ($\gamma$-$Fe_2O_3$). Maruthamuthu and Huie (1995) also investigated the photooxidation of iron-carboxyl complexes of haloacetic acid and came to the conclusion that the reaction rate of TFA was "too slow to measure." They also suggest that the mechanism of the reaction is mediated by the generation of a $OH\bullet$ radical that attacks the haloacetic acid. A review by Boutonnet et al. (1999) summarized additional experiments that were presented as gray literature reports, but the main conclusion from these reports is that TFA is stable or degrades very slowly under most environmental conditions.

The other environmentally relevant process by which TFA could be degraded would be reduction in anaerobic environments. An early experiment suggested that TFA could be reductively degraded by anoxic sediment samples (Visscher et al., 1994) although subsequent research using samples from similar areas were unable to replicate the result (Emptage et al., 1997). TFA was degraded in an anaerobic reactor at 35°C with ethanol added to it, which proves reductive dehalogenation can occur as a cometabolism process under certain conditions (Kim et al., 2000). More recently, a 33-week incubation of anaerobic sludge showed no degradation of TFA (Ochoa-Herrera et al., 2016). Another study did not detect TFA degradation in vernal pool water sealed in glass jars for 133 days during which the samples became anaerobic (Cahill et al., 2001). The conflicting reports on the reduction of TFA under anaerobic conditions give rise to the conclusion that the reductive dehalogenation of TFA may be possible, but the circumstances under which it can occur are probably limited.

### 9.4.3 Toxicity

The last physicochemical property of TFA that is noteworthy is its toxicity, or more accurately, the lack thereof. The protonated form of TFA is a strong acid that can cause burns due to its acidic character in a fashion similar to that of hydrochloric acid (Blake et al., 1969). However, protonated TFA is effectively absent in the environment due to its low $pK_a$ with the possible exception of TFA generated in the atmosphere as a vapor that had not yet partitioned into

condensed phases, such as rain and aerosols. Therefore, the assessments of
TFA toxicity were generally conducted with the ionic form of the molecule.
More detailed reviews of the toxicology of TFA can be found in Boutonnet
et al. (1999) and Solomon et al. (2016).

The toxicity of TFA in animals is low (Table 9.2). The lack of reactivity
of TFA that makes it stable in the environment also makes it unreactive in
the body. In mammals, no toxic effects of the ionic form of TFA have been
reported at 5,000 mg/kg (Blake et al., 1969) and 2.1 mmol/kg (=237 mg/kg)
(Fraser and Kaminsky, 1988). TFA is predominately excreted from mammals
in the urine with half-lives ranging from 16 to 61 hours (Holaday, 1977). Due
to the relatively short half-life of TFA, it will not significantly accumulate in
animals. TFA appears to bind to carrier proteins, such as albumin, and other
macromolecules, which may be part of the reason its elimination was slower
than expected. It has been suggested that TFA may form a glucuronide conjugate in the liver only to be excreted in the bile and reabsorbed in the intestine.
This potential enterohepatic circulation might be another reason for the slow
elimination kinetics of TFA (Boutonnet et al., 1999) compared to other small,
unreactive ions. Lastly, TFA is not mutagenic (Waskell, 1978). In essence, TFA
behaves in animals as a small ionic molecule that is relatively inert and easily
excreted in the urine.

The toxicity of TFA in aquatic systems was also fairly low (Table 9.2). The
estimated no observable effect concentration (NOEC) in aquatic microcosms
was 11.8 mg TFA/L (Solomon et al., 2016). A study of TFA toxicity toward a
variety of aquatic organisms, such as algae, diatoms, vascular plants, invertebrates, and fish, showed the effective concentration-50 ($EC_{50}$) of all organisms
to be 112 mg/L or higher with the exception of one algae, namely, *Selenastrum
capricornutum* (Berends et al., 1999). This sensitive species had an $EC_{50}$ of
approximately 1.2 mg/L and a NOEC of 0.12 mg/L. The authors conclude
that a concentration of 0.10 mg/L would be safe for the aquatic ecosystem
(Berends et al., 1999). Another microcosm study showed that a mixture of 10
mg/L of trichloroacetic acetic acid (TCA) and TFA did not produce long-term
effects on two aquatic macrophytes (Hanson et al., 2002). Additional studies
of aquatic macrophytes showed the $EC_{50}$ for TFA ranged from 221.1 to 10,000
mg/L for three macrophyte species using a variety of indices for potential
impacts on plants (Hanson and Solomon, 2004). Overall, the average NOEC
for aquatic plants and algae was approximately 520 mg/L (Table 9.2), which
makes aquatic plants and algae rather resistant to TFA although some excep-
tions occur such as that for *Selenastrum capricornutum*.

The toxicity of TFA in terrestrial plants was expected to be higher than
aquatic plants since terrestrial plants can bioaccumulate TFA. Basically, terrestrial plants can absorb and translocate water and TFA together, but when
the water evaporates from the plant, the TFA is left behind and concentrates
in the leaves (Rollins et al., 1989). The bioconcentration factors (BCF) for TFA
in terrestrial plants were: from 4.9 to 43 over about a 35-day period (data summarized in Boutonnet et al. (1999)); 26 to 295 (Benesch et al., 2002); and 18 to 61

**TABLE 9.2**

Summary of Toxic Effects of TFA on Organisms

| Species | Type of Organism | No Observed Effects (NOEL or NOEC) | 50% Effect (TD$_{50}$ or EC$_{50}$) | Reference |
|---|---|---|---|---|
| *Animals:* | | | | |
| Mice | Mammal | 5,000 mg/kg | | Blake et al. (1969) |
| Rats | Mammal | 237 mg/kg | | Fraser and Kaminsky (1988) |
| *Danio rerio* | Fish | 1,200 mg/L | >1,200 mg/L | Berends et al. (1999) |
| *Daphnia magna* | Aquatic invertebrate | 1,200 mg/L | >1,200 mg/L | Berends et al. (1999) |
| *Aquatic Plants and Algae:* | | | | |
| *Lemna gibba* | Aquatic macrophyte | 300 mg/L | 1,100 mg/L | Berends et al. (1999) |
| *Lemna gibba* | Aquatic macrophyte | | 618.3->3,000 mg/L | Hanson and Solomon (2004) |
| *Myriophyllum spicatum* | Aquatic macrophyte | | 312.9->10,000 mg/L | Hanson and Solomon (2004) |
| *Myriophyllum sibiricum* | Aquatic macrophyte | | 340.7 to >10,000 mg/L | Hanson and Solomon (2004) |
| *Selenastrum capricornutum* | Green alga | 0.12 and <0.36 mg/L | >1.2 and 4.8 mg/L | Berends et al. (1999) |
| *Chlorella vulgaris* | Green alga | 1,200 mg/L | >1,200 mg/L | Berends et al. (1999) |
| *Scenedesmus subspicatus* | Green alga | | >120 mg/L | Berends et al. (1999) |
| *Chlamydomonas reinhardtii* | Green alga | 120 mg/L | >120 mg/L | Berends et al. (1999) |
| *Dunaliella tertiolecta* | Green alga | <124 mg/L | >125 mg/L | Berends et al. (1999) |
| *Euglena gracilis* | Green alga | 112 mg/L | >112 mg/L | Berends et al. (1999) |
| *Phaeodactylum tricornutum* | Diatom | 117 mg/L | >117 mg/L | Berends et al. (1999) |
| *Navicula pelliculosa* | Diatom | 600 mg/L | 1,200 mg/L | Berends et al. (1999) |
| *Skeletonema costatum* | Diatom | 2,400 mg/L | >2,400 mg/L | Berends et al. (1999) |
| *Anabaena flos-aquae* | Blue-green alga | 600 mg/L | 2,400 mg/L | Berends et al. (1999) |
| *Microcystis aeruginosa* | Blue-green alga | 117 mg/L | >117 mg/L | Berends et al. (1999) |
| Average NOEC for aquatic plants and algae | | 520 mg/L[f] | | |

(Continued)

**TABLE 9.2 (*Continued*)**

Summary of Toxic Effects of TFA on Organisms

| Species | Type of Organism | No Observed Effects (NOEL or NOEC) | 50% Effect (TD$_{50}$ or EC$_{50}$) | Reference |
|---|---|---|---|---|
| *Terrestrial Plants* | | | | |
| *Deschampsia elongata* | Terrestrial plant | >1 mg/L[b] | | Benesch et al. (2002) |
| *Lasthenia californica* | Terrestrial plant | >1 mg/L[b] | | Benesch et al. (2002) |
| *Oryza sativa* (rice) | Terrestrial plant | >1 mg/L[b] | | Benesch et al. (2002) |
| *Pinus ponderosa* | Terrestrial plant | >10 mg/L[c] | | Benesch and Gustin (2002) |
| *Phaseolus vulgaris* (common bean) | Terrestrial plant | >2.5mg/L; LOEL=10 mg/L[b,d] | | Smit et al. (2009) |
| *Vigna radiata* (mung bean) | Terrestrial plant | 1 mg/kg of soil[b] | | Boutonnet et al. (1999)[e] |
| *Zea mays* (corn) | Terrestrial plant | >2.5mg/L; LOEL=10 mg/L[b,d] | | Smit et al. (2009) |
| *Triticum aestivum* (wheat) | Terrestrial plant | 1-32 mg/L[b] 50 mg/L[c] | | Boutonnet et al. (1999)[e] |
| *Plantago major* | Terrestrial plant | 32 mg/L[b] | | Boutonnet et al. (1999)[e] |
| *Helianthus annuus* (sunflower) | Terrestrial plant | <1 mg/kg soil[b] 100 mg/L[c] | | Boutonnet et al. (1999)[e] |
| *Glycine max* (soybean) | Terrestrial plant | 1 mg/L[b] 10 mg/L[c] | | Boutonnet et al. (1999)[e] |
| Average NOEC for terrestrial plants | | 7.2 mg/L[b,f] | | |

[a] Test was a mixture of TFA and TCA that showed no long-term effects. There were transitory effects in the first few days.

[b] Root uptake route of exposure.

[c] Foliar uptake route of exposure.

[d] Based on biomass production. However, other indices of plant performance showed effects as low as 0.625 mg/L.

[e] Data summarized in Boutonnet et al. (1999) from gray literature reports.

[f] The average NOEL calculations used the minimum possible values such that treated values reported as ">1" as 1.

(Cahill et al., 2001). The NOEC was determined for a series of terrestrial plants with crop species being the most common test organisms. The NOEC for the root uptake mechanism ranged from approximately 1 to 32 mg/L (Table 9.2). The average NOEC over several species was 7.2 mg/L although this is an underestimate since a large number of the toxicology data are reported as lower limits. This is considerably lower than the estimated NOEC for aquatic plants at 520 mg/L since the aquatic plants lacked evaporative bioconcentration. Given the affinity of TFA for water, it is expected that aquatic plants are more likely be exposed to elevated TFA concentrations. However, terrestrial plants associated with water systems, and terminal water systems in particular, may suffer from both elevated concentrations and bioaccumulation in their tissues.

## 9.5 Environmental Concentrations of TFA

TFA is ubiquitous in the environment with the possible exception of pre-industrial age fresh water. It is found in deep ocean water to remote Polar Regions although it is more concentrated in urban areas where there are more potential sources. However, TFA generated from HFC degradation has changed over time as the usage of the HFCs has increased, further complicating matters. This has changed the deposition of TFA on a global basis. There are also specific hot spots near point sources of the precursors to TFA. For this analysis, the reported environmental TFA concentrations will be assigned to specific conditions that they represent, namely, oceanic, remote hemispheric background, populated regions, terminal waterbodies, and point sources of TFA.

### 9.5.1 Oceans

The discovery of TFA in the oceans was not anticipated since TFA was originally considered to be an anthropogenic pollutant. However, numerous studies by multiple research groups have verified the presence of TFA in the oceans. In general, TFA concentrations range from 1 to 250 ng/L (Table 9.3) although there seems to be a trend to higher concentrations in the Atlantic and Arctic Oceans compared to the Pacific Ocean (Scott et al., 2005). Many of these water samples were from deep ocean currents whose water was dated to preindustrial times, thus suggesting that there is a natural source of TFA that has accumulated in the oceans for millennia. Geothermal vents were shown to have elevated concentrations of TFA over them, so this identifies at least one of the natural sources of TFA in the oceans (Scott et al., 2005). The uniformity of TFA concentrations in the water column also suggests that TFA is very stable due to the slow mixing of the ocean waters (Frank et al., 2002).

**TABLE 9.3**

Oceanic Concentrations (ng/L) of TFA

| Location | Year | Concentrations | Reference |
|---|---|---|---|
| Arctic (Canada Basin) | 2005 | 34–181 | Scott et al. (2005) |
| Arctic (Eastern) | 2005 | 8–170 | Scott et al. (2005) |
| Arctic (Norway) | 1996 | 64 and 65 | Berg et al. (2000) |
| Atlantic (North, deep water) | 2005 | 28–190 | Scott et al. (2005) |
| Atlantic (France) | 1995 | 250 | Frank et al. (1996) |
| Atlantic (Ireland) | 1995 | 70 | Frank et al. (1996) |
| Atlantic (North Sea) | 1995 | 90 | Frank et al. (1996) |
| Atlantic (Baltic Sea) | 1995 | 40 | Frank et al. (1996) |
| Atlantic (Middle, surface and deep water) | 1998 | 190–210 | Frank et al. (2002) |
| Atlantic (South, deep water) | 2005 | 64–200 | Scott et al. (2005) |
| Pacific (North, excluding geothermal vents) | 2005 | 1–80 | Scott et al. (2005) |
| Pacific (Vietnam) | 1996 | 53 and 66 | Berg et al. (2000) |
| Pacific (China, coastal and beach) | 2001 | 4–190 | Zhang et al. (2005) |
| Pacific (South, deep water) | 2005 | 1–150 | Scott et al. (2005) |
| Southern Ocean (surface and deep water) | 1998 | 165–220 | Frank et al. (2002) |

The oceans can be considered the ultimate repository for TFA with the exception of some relatively rare terminal lakes. TFA's stability and ionic character will make it behave in a similar fashion as other ionic salts.

### 9.5.2 Hemispheric Background Concentrations

TFA was expected to be globally distributed due to the long atmospheric lifetime of some of the HFC precursors like HFC-134a. Surprisingly, there are relatively few measurements of TFA in remote background sites. Rivers and lakes in Alaska, Yukon Territory, and British Colombia had median concentrations around 21 ng/L (Cahill and Seiber, 2000). An ice core from Antarctica showed that the average TFA concentration was about 25 ng/L and old ice from a glacier in northern Sweden had a concentration of about 5 ng/L (Von Sydow et al., 2000). However, another study did not detect TFA in an ice core from the Greenland Summit that was dated to be about 4,000 years old (Nielsen et al., 2001). Snow samples collected between 1994 and 1997 in remote areas generally showed low concentrations of TFA between 1 and 110 ng/L (Von Sydow et al., 2000). Saturna Island, which is an island in British Colombia, Canada, that was extensively influenced by the Pacific Ocean, had median concentrations in rainfall of 21 ng/L (range 3–31 ng/L) (Scott et al., 2005), 29 ng/L (range <0.1–170) (Scott et al., 2000), and 5–140 ng/L (Scott et al., 2006). Two remote lakes in western Canada, namely, Loon Lake and Great Slave Lake, had concentrations <0.5 ng/L and between <0.5 and 10 ng/L, respectively (Scott et al., 2000). Rain at Snare Rapids, Northwest Territories,

Canada, had concentrations less than 0.5 ng/L (Scott et al., 2000). Streams along coastal Northern California, which represent Pacific air masses, had concentrations averaging 24.3 ng/L (Cahill and Seiber, 2000). Overall, these data suggest that TFA is globally distributed at low concentrations as would be expected with a compound degrading from HFC precursors. However, most of these sites were sampled before 2001 and there are little recent data available on TFA in remote areas, which are needed given the increase in the atmospheric concentrations of HFC-134a.

### 9.5.3 Populated Regions

Considerable research has been conducted investigating TFA concentrations in populated regions ranging from rural locations to large industrial cities. This summary will focus on aqueous media, such as rain, fogwater, rivers, and lakes, since there are more data available for comparisons. Other media, such as air, soil, and plants (Cahill et al., 2001; Chen et al., 2018; Martin et al., 2003; Scott et al., 2005; Tian et al., 2018; Wu et al., 2014; Xie et al., 2020; Zhang et al., 2018), have also been sampled, but at a considerably lower frequency. There have also been extensive modeling efforts to predict TFA concentrations resulting from precursor compounds such as HFC-134a and HFO-1234yf (Henne et al., 2012; Kazil et al., 2014; Kotamarthi et al., 1998; Luecken et al., 2010; Wang et al., 2018; Wu et al., 2014).

A summary of TFA in surface waters of populated areas (Table 9.4) shows that most TFA concentrations are below 300 ng/L and effectively all measurements are below 1,000 ng/L. Higher concentrations were detected in highly industrialized and/or populated areas. The high concentrations in Germany were largely attributed to fluorochemical manufacturing in the region (Janda et al., 2019; Scheurer et al., 2017). High concentrations of TFA were also observed in Beijing, China, which was expected due to the large population in the area. However, these samples also provided a time series on how these concentrations changed over time since the same sites were sampled in 2001 and 2012. During the decade between sample collections, the concentrations in the surface waters increased approximately 17-fold (Zhai et al., 2015). This increase could be attributed to many potential factors, such as increased industrialization in China, greater atmospheric HFC concentrations, and/or greater consumer use of fluorinated chemicals in the region. To put these surface water concentrations into perspective, the most sensitive aquatic organism to TFA was an alga with a NOEC of 0.12 mg/L, which equates to 120,000 ng/L. Since the surface water concentrations are typically less than 1,000 ng/L, the concentrations are less than 100-fold the toxic effect level.

A review of the literature on surface water concentrations of TFA showed a diminished publication rate after approximately 2006 with the exception of sampling near point sources. This can largely be attributed to the increased attention to longer chain and more toxic fluorinated acids like PFOS and

**TABLE 9.4**

Concentrations of TFA (ng/L) in Rivers and Lakes in Populated Regions

| Location | Media | Year | Concentrations (ng/L) | Ref. |
|---|---|---|---|---|
| Canada/USA (Great Lakes) | Lakes | 2000 | <0.5–150 | Scott et al. (2000) |
| Canada/USA (Lake Superior) | Lakes | 1998 | **18** (mean) | Scott et al. (2002) |
| Canada/USA (Lake Ontario) | Lakes | 1998 | **150** (mean) | Scott et al. (2002) |
| Canada/USA (Detroit River) | Rivers | 1998 | 51–99 | Scott et al. (2002) |
| China (multiple provinces) | Rivers/lakes | 2001 | 6.8–221 | Zhang et al. (2005) |
| China (Beijing)[a] | lakes | 2012 | **643**±169 | Zhai et al. (2015) |
| Germany | Rivers/lakes | 1995 | 60–650 | Frank et al. (1996) |
| Germany | Rivers/lakes | 2017 | 480–1,200 with outliers as high as 17,000 | Janda et al. (2019) |
| Israel | Rivers/lakes | 1995 | 200–250 | Frank et al. (1996) |
| Switzerland | Rivers | 1997 | **87** (range 12–328) | Berg et al. (2000) |
| Switzerland | Lakes | 1997 | **119** (range 37–360) | Berg et al. (2000) |
| USA (California, urban) | Streams | 1998 | **95.4** and **144** (means) | Cahill and Seiber (2000) |
| USA (California, rural) | Streams | 1998 | **21.5** and **24.1** and **27.8** (means) | Cahill and Seiber (2000) |
| USA (Nevada) | Rivers | 1997 | 12.8–154 | Wujcik et al. (1999) |

Values in bold denote a reported mean value.

[a] This study resampled sites sampled in 2001 and found a 17-fold increase in TFA concentrations over a decade.

PFOA. However, additional sampling of both remote and "typical" surface waters is warranted in order to understand the impacts of the elevated concentrations of HFC-134a (Figure 9.3) in the atmosphere are having on TFA formation on a global scale.

Other frequently sampled media for TFA are precipitation and fogwater (Table 9.5). The observed concentrations were similar in magnitude as the surface waters with almost all of the concentrations falling below 1,000 ng/L and most falling below 500 ng/L. The precipitation concentrations have some higher concentrations due to low volume rain events where TFA is concentrated into a small volume of precipitation. There is a clear "wash out" effect where larger rain events have lower concentrations since most of the TFA was scrubbed from the atmosphere during the initial part of the rain event and the subsequent rain simply diluted the sample (Berg et al., 2000; Wujcik et al., 1998). Only a few studies reported fogwater concentrations of TFA and it was found that TFA concentrations in fogwater were considerably higher than rainfall in the same region (Rompp et al., 2001; Wujcik et al., 1998; Zehavi and Seiber, 1996). This was expected since fog has a low volume of water and a longer residence time in which to accumulate TFA.

**TABLE 9.5**

Concentrations of TFA (ng/L) in Precipitation in Populated Regions

| Location | Media | Year | Concentrations (ng/L) | Ref. |
|---|---|---|---|---|
| Canada (Ontario, urban) | Precipitation | 2004 | 87–270 | Scott et al. (2006) |
| Canada (Ontario, remote) | Precipitation | 1997 | **48** (range 4–120) | Scott et al. (2000); |
| | | 1999 | **94** (range 12–350) | Scott et al. (2006); |
| | | 2002 | 8–220 | Scott et al. (2005) |
| Canada (Chapias, rural) | Precipitation | 1997 | **24** (range <0.1–69) | Scott et al. (2000); |
| | | 1999 | **34** (range <0.5–54) | Scott et al. (2006); |
| | | 2001 | 140 | Scott et al. (2005) |
| Chile (urban) | Precipitation | 1999 | **14** (range 6–85) | Scott et al. (2005) |
| Chile (rural) | Precipitation | 1999 | **12** (range 5–12) | Scott et al. (2005) |
| China (Guangdong province) | Precipitation | 2008 | **225** (range 45.8–974) | Wang et al. (2014) |
| China (Beijing) | Precipitation | 2001 | 25–220 | Zhang et al. (2005) |
| China (Beijing)[a] | Precipitation | 2012 | **282±68** | Zhai et al. (2015) |
| China (multiple provinces) | Precipitation | 2018 | **150** (range 8.8–1,800) | Chen et al. (2019) |
| Germany | Precipitation | 1999 | **100** median (20–930) **230** median (20–420) | Rompp et al. (2001) |
| Germany (Bavaria) | Fogwater | 1999 | **230** (range 20–1,900) | Rompp et al. (2001) |
| Ireland (Mace Head) | Precipitation | 1996 | 2–92 | Von Sydow et al. (2000) |
| Japan (Tsukuba and Kawaguchi) | Precipitation | 2007 | 39.3–75.9 | Taniyasu et al. (2008) |
| Malawi (Senga Bay) | Precipitation | 1999 | **8.8** (range 4–15) | Scott et al. (2005) |
| Poland (Gdansk) | Precipitation | 1996 | 26–1100 | Von Sydow et al. (2000) |
| Russia (near Lake Baikal) | Precipitation | 1996 | 30–215 | Berg et al. (2000) |
| Switzerland | Precipitation | 1997 | **151** (range <3–1,550) | Berg et al. (2000) |
| USA (California and Nevada) | Precipitation | 1994 | 31–90 | Zehavi and Seiber (1996) |
| USA (California) | Fogwater | 1994 | > 208; >280 and 279 | Zehavi and Seiber (1996) |
| USA (California, urban) | Precipitation | 1999 | 27–302 | Cahill et al. (2001) |
| USA (California and Nevada) | Precipitation | 1997 | **46.6** and **63.9** and **136** (means) | Wujcik et al. (1999) |
| USA (California) | Fogwater | 1997 | **689** and **723** (means) | Wujcik et al. (1999) |
| USA (New York, rural) | Precipitation | 1999 | 3–360 | Scott et al. (2006) |
| USA (Maryland, near urban) | Precipitation | 1999 | 21–620 | Scott et al. (2006) |
| USA (Delaware, near urban) | Precipitation | 1999 | 10–2,400 | Scott et al. (2006) |
| USA (Vermont, rural) | Precipitation | 1999 | 8–250 | Scott et al. (2006) |
| Vietnam (Hanoi) | Precipitation | 1996 | 4–150 | Berg et al. (2000) |

Values in bold denote a reported mean value.

Just like surface water measurements, precipitation measurements have declined in recent years, so the majority of the data available were collected prior to 2001. This makes it difficult to determine how the concentrations are changing over time. One study in Beijing, China, showed that precipitation concentrations of TFA have increased dramatically over a decade (Zhai et al., 2015). Precipitation monitoring sites in Canada have also seen increasing concentrations between 1997 and 1999, although more recent data (2001 and 2002) are too sparse to make any conclusions. It would be expected that TFA concentrations in precipitation would be higher now due to the elevated concentrations of HFC-134a and the introduction of HFO-1234yf, which forms more TFA on a mole basis than HFC-134a.

TFA also appears in groundwater samples where the water arises from recent infiltration. TFA behaves in a similar fashion as chloride or bromide ions in the soil (Richey et al., 1997): so it is fairly easily percolated into groundwater from atmospheric or aquatic sources. Shallow groundwater samples in China had similar concentrations of TFA as the surface water samples and rain samples (Chen et al., 2018), which also demonstrates the lack of permanent losses of TFA passing through soil. Similar TFA concentrations between surface waters and young spring water was also observed in Germany (Jordan and Frank, 1999). TFA frequently appears in well water of recent origins (Xie et al., 2020). In Switzerland, it was estimated that approximately 62% of the TFA in the region was lost to groundwater while the remaining 38% was exported in surface waters (Berg et al., 2000). Overall, TFA appears to be easily transported to groundwater which is often used as a source of drinking water.

### 9.5.4 Terminal Waterbodies

The main concern with TFA is that its stability would allow it to accumulate in terminal waterbodies that lack a water outflow. Precipitation and surface water inputs would add TFA to these lakes. While the water can evaporate from these lakes, the TFA cannot and so it remains behind, as a result of evapoconcentration. After a period of years, the TFA concentrations would rise to the concentrations that could impact aquatic organisms (Russell et al., 2012; Tromp et al., 1995). These types of lakes tend to occur in arid regions that have low precipitation and high evaporation rates. Despite the concern over these terminal waterbodies, very little sampling has been conducted in these ecosystems and all of the data predate the year 2000 (Table 9.6). The Dead Sea in Israel had concentrations that were comparable to surface waters nearby (Frank et al., 1996). In contrast, three lakes sampled in the United States had elevated concentrations compared to the rivers flowing into them (Wujcik et al., 1999). This implies that the TFA was building up in the ecosystem. It was estimated that the observed lake concentrations represented between 4.2 and 13 years of inputs at the current rate. This sampling represents a time period (1997) where HFC-134a emissions and atmospheric concentrations were still relatively low. None of these lakes had any heavy industry in their

**TABLE 9.6**

Concentrations of TFA (ng/L) in Terminal Waterbodies

| Location | | Year | Concentrations | Reference |
|---|---|---|---|---|
| Israel | Dead Sea | 1995 | 250 | Frank et al. (1996) |
| USA (California) | Vernal pools | 1999 | Up to 10,000 during evaporation | Cahill et al. (2001) |
| USA (Nevada) | Pyramid Lake[a] | 1997 | 79 (range 77.1–95.1) | Wujcik et al. (1999) |
| USA (California) | Mono Lake | 1997 | 192 (range 186–227) | Wujcik et al. (1999) |
| USA (Nevada) | Stillwater NWR | 1997 | 432.5 (range 314–472) | Wujcik et al. (1999) |

[a] Two measurements in 1994 gave concentrations of approximately 40,000 ng/L (Zehavi and Seiber, 1996), but more extensive sampling in 1997 resulted a much lower concentration reported here that are more credible and more consistent with both the inflowing river and other nearby terminal lakes.

watersheds and the wastewater effluent from Reno, NV has no appreciable impact on the river concentrations leading to Pyramid Lake (Wujcik et al., 1999). Therefore, the main input was expected to be atmospheric and HFCs were the suspected cause. If this was the case, then the lakes would have a large volume of TFA-free water prior to HFC introduction. Once atmospheric sources of TFA started adding TFA to the lake, the concentrations would be diluted by the large lake volume. This would result in the smaller lakes, like Stillwater National Wildlife Refuge, responding faster to the TFA input than the larger lakes like Pyramid Lake. This was the observed trend with Stillwater National Wildlife Refuge having the highest TFA concentrations and the largest lake having the lowest concentrations (Wujcik et al., 1999). Once again, this study was conducted near the beginning of the HFC adoption and the concentrations have undoubtedly changed since then. Thus, these lakes are in a desperate need of a resampling effort to see how the concentrations have changed over 20 years.

The other terminal waterbodies sampled were vernal pools in California, which was the exact ecosystem that was hypothesized to be vulnerable to TFA (Tromp et al., 1995). Vernal pools are seasonal wetlands that fill with rain during the winter and then dry out during the summer. In California, they also have several rare and endangered species. The water, soil, and plants growing in vernal pools were sampled over a couple of years and the results were exactly as expected (Cahill et al., 2001). The first water that collected in the pool had relatively high concentrations of TFA, which indicated that it was solubilizing TFA left in the soil from the prior year. As the pools filled, the concentrations went down by simple dilution. Once the rains stopped, the pools started to evaporate and the TFA concentrations increased indicating that the TFA was being concentrated in a smaller volume of water. The highest concentrations, up to 10,000 ng/L, were observed in the last water in the pools before they dried up (Cahill et al., 2001). These concentrations were the highest recorded concentrations at the time and are still very high compared to other surface waters excluding waters near point sources of TFA. It also clearly showed the year-to-year carryover of TFA as predicted. Furthermore, the TFA was being taken up into plants and concentrating even more in the plant tissues (Cahill et al., 2001). Once again, this study was conducted fairly early in the HFC adoption and atmospheric formation of TFA is expected to be higher now. With the carryover of TFA between seasons and the increased deposition of TFA, the concentrations are likely higher now and warrant additional research.

### 9.5.5 Point Sources of TFA

The last main set of reported environmental TFA data are concentrations near point sources such as fluorochemical manufacturing facilities, landfills, and wastewater treatment plants (Table 9.7). The three sites with the highest concentration of TFA were wastewater from fluorochemical facilities or in rivers downstream of such facilities (Berg et al., 2000; Chen et al., 2018; Scheurer et al., 2017). Some of these facilities manufacture TFA as a product or use it as an intermediate chemical. Some of this chemical is lost in the process and ends up being discharged in the wastewater from the facility. It is worth noting that the highest concentration reported (207,000 ng/L) (Berg et al., 2000) is

**TABLE 9.7**

Concentrations of TFA (ng/L) Near Point Sources

| Location | | Year | Concentrations | Reference |
|---|---|---|---|---|
| China (Liaoning Province) | Rain near fluorochemical plants | 2016 | 580–800 | Chen et al. (2018) |
| China (Liaoning Province) | Surface water near fluorochemical plants | 2016 | 670–35,000 | Chen et al. (2018) |
| China (Shandong Province) | Rivers in fluorochemical plant region | 2016 | 932±56; 1,014±79; 588±72 | Xie et al. (2020) |
| Germany | River downstream of fluorochemical plant | 2017 | 5,400–140,000 | Scheurer et al. (2017) |
| Germany | Wastewater treatment plant effluents | 2017 | 570 to 1,300; 4,200±880 | Scheurer et al. (2017) |
| Sweden | Landfill leachate | 2018 | <34–6,900 | Bjornsdotter et al. (2019) |
| Sweden | Fire-fighting training site | 2018 | <34–14,000 | Bjornsdotter et al. (2019) |
| Sweden | Downstream of a hazardous materials waste facility | 2018 | <34–2,700 | Bjornsdotter et al. (2019) |
| Switzerland | Industrial wastewater | 1997 | 47,300 (range <100–207,000) | Berg et al. (2000) |
| Switzerland | Communal wastewater | 1997 | 230 (range 90–600) | Berg et al. (2000) |
| Switzerland | Swimming pools | 1997 | 4,800 (range 4,100–5,700) | Berg et al. (2000) |

above the concentration that is expected to cause harm to the most sensitive algae (120,000 ng/L) (Berends et al., 1999) although this was a grab sample from the effluent of an industrial wastewater treatment plant. However, TFA concentrations near fluorochemical facilities routinely range from 1,000 to 10,000 ng/L, which are concentrations that start to become worrisome from a toxicology perspective since it leaves less of a safety margin before toxic effects could be reached. TFA also arises from wastewater treatment facilities serving residential populations, but the concentrations are much lower and the suspected source might be from pharmaceuticals and domestic fluoro-chemical use (Scheurer et al., 2017). Other point sources of TFA include land-fills (Bjornsdotter et al., 2019; Tian et al., 2018) and firefighting training sites where fluorinated surfactants, such as aqueous film-forming foams (AFFF), were utilized (Bjornsdotter et al., 2019).

## 9.6 Conclusions

TFA is a ubiquitous halogenated acid that is very stable and likely to per-sist in the environment for considerable time. It is almost exclusively found in the ionic form in the environment which makes it highly water soluble and nonvolatile. In many ways, it behaves in a similar fashion as other small ions such as chloride or bromide ions. These characteristics of the chemical predispose it to partition into water and stay in the aqueous system. Unlike the larger perfluorinated acids (e.g., PFOS and PFOA), it does not bioaccu-mulate in animals and it has relatively low toxicity. Its main toxicity appears to be toward plants; terrestrial plants can bioaccumulate TFA in their above ground biomass. The fate of most of the TFA generated is to be washed out of the atmosphere and flow to the ocean in rivers and streams where it will join a large pool of natural TFA whose source is still not completely clear. The main concern regarding TFA is in water systems in arid areas that lack a water outflow, so the TFA may accumulate over years to reach toxic concen-trations. While current concentrations are still below toxic concentrations, they need to be monitored to ensure that environmental impacts do not develop. If negative impacts develop in these ecosystems, then the impacts are expected to last for a very long time due to the stability of TFA.

## References

Andersen, M.P.S., Nielsen, O.J., Karpichev, B., Wallington, T.J., and Sander, S.P. (2012). Atmospheric chemistry of Isoflurane, Desflurane, and Sevoflurane: Kinetics and mechanisms of reactions with chlorine atoms and OH radicals and global warming potentials. *J. Phys. Chem. A*, 116(24), 5806–5820. Doi: 10.1021/jp2077598.

Benesch, J.A., and Gustin, M.S. (2002). Uptake of trifluoroacetate by Pinus ponderosa via atmospheric pathway. *Atmos. Environ.* 36(7), 1233–1235, Article Pii s1352-2310(01)00562-3. Doi: 10.1016/s1352-2310(01)00562-3.

Benesch, J.A., Gustin, M.S., Cramer, G.R., and Cahill, T.M. (2002). Investigation of effects of trifluoroacetate on vernal pool ecosystems. *Environ. Toxicol. Chem.* 21 (3), 640–647. Doi: 10.1897/1551-5028(2002)021<0640:Ioeoto>2.0.Co;2.

Berends, A.G., Boutonnet, J.C., de Rooij, C.G., and Thompson, R.S. (1999). Toxicity of trifluoroacetate to aquatic organisms. *Environ. Toxicol. Chem.* 18(5), 1053–1059. Doi: 10.1002/etc.5620180533.

Berg, M., Muller, S.R., Muhlemann, J., Wiedmer, A., and Schwarzenbach, R.P. (2000). Concentrations and mass fluxes of chloroacetic acids and trifluoroacetic acid in rain and natural waters in Switzerland. *Environ. Sci. Technol.* 34(13), 2675–2683. Doi: 10.1021/es990855f.

Bjornsdotter, M.K., Yeung, L.W.Y., Karrman, A., and Jogsten, I.E. (2019). Ultra-short-chain perfluoroalkyl acids including trifluoromethane sulfonic acid in water connected to known and suspected point sources in Sweden. *Environ. Sci. Technol.* 53(19), 11093–11101. Doi: 10.1021/acs.est.9b02211.

Blake, D.A., Cascorbi, H.F., Rozman, R.S., and Meyer, F.J. (1969). Animal toxicity of 2,2,2-trifluoroethanol. *Toxicol. Appl. Pharmacol.* 15(1), 83. Doi: 10.1016/0041-008x(69)90135-5.

Booten, C., Nicholson, S., Mann, M., and Abdelaziz, O. (2020). *Refrigerants: Market Trends and Supply Chain Assessment* (NREL/TP-5500-70207). https://www.nrel.gov/docs/fy20osti/70207.pdf.

Boutonnet, J.C., Bingham, P., Calamari, D., de Rooij, C., Franklin, J., Kawano, T. et al., (1999). Environmental risk assessment of trifluoroacetic acid. *Human Ecol. Risk Assess.* 5(1), 59–124. Doi: 10.1080/10807039991289644.

Bowden, D.J., Clegg, S.L., and Brimblecombe, P. (1996). The Henry's law constant of trifluoroacetic acid and its partitioning into liquid water in the atmosphere. *Chemosphere* 32(2), 405–420. Doi: 10.1016/0045-6535(95)00330-4.

Cahill, T.M., and Seiber, J.N. (2000). Regional distribution of trifluoroacetate in surface waters downwind of urban areas in Northern California. USA. *Environ. Sci. Technol.* 34(14), 2909–2912. Doi: 10.1021/es991435t

Cahill, T.M., Thomas, C.M., Schwarzbach, S.E., and Seiber, J.N. (2001). Accumulation of trifluoroacetate in seasonal wetlands in California. *Environ. Sci. Technol.* 35(5), 820–825. Doi: 10.1021/es0013982.

Carr, S., Treacy, J.J., Sidebottom, H.W., Connell, R.K., Canosamas, C.E., Wayne, R.P., and Franklin, J. (1994). Kinetics and mechanisms for the reaction of hydroxyl radicals with trifluoroacetic acid under atmospheric conditions. *Chem. Phys. Lett.* 227(1–2), 39–44. Doi: 10.1016/0009-2614(94)00802-7.

Chen, H., Yao, Y.M., Zhao, Z., Wang, Y., Wang, Q., Ren, C., Wang, B., Sun, H.W., Alder, A.C., and Kannan, K. (2018). Multimedia distribution and transfer of per- and polyfluoroalkyl substances (PFASs) surrounding two fluorochemical manufacturing facilities in fuxin, China. *Environ. Sci. Technol.* 52(15), 8263–8271. Doi: 10.1021/acs.est.8b00544.

Chen, H., Zhang, L., Li, M.Q., Yao, Y.M., Zhao, Z., Munoz, G., and Sun, H.W. (2019). Per- and polyfluoroalkyl substances (PFASs) in precipitation from mainland China: Contributions of unknown precursors and short-chain (C2-C3) perfluoroalkyl carboxylic acids. *Water Res.* 153, 169–177. Doi: 10.1016/j.watres.2019.01.019.

Ellis, D.A., Hanson, M.L., Sibley, P.K., Shahid, T., Fineberg, N.A., Solomon, K.R., Muir, D.C.G., and Mabury, S.A. (2001a). The fate and persistence of trifluoroacetic and chloroacetic acids in pond waters. *Chemosphere*, 42(3), 309–318. Doi: 10.1016/s0045-6535(00)00066-7.

Ellis, D.A., and Mabury, S.A. (2000). The aqueous photolysis of TFM and related trifluoromethylphenols. An alternate source of trifluoroacetic acid in the environment. *Environ. Sci. Technol.* 34(4), 632–637. Doi: 10.1021/es990422c.

Ellis, D.A., Mabury, S.A., Martin, J.W., and Muir, D.C.G. (2001b). Thermolysis of fluoropolymers as a potential source of halogenated organic acids in the environment. *Nature*, 412(6844), 321–324. Doi: 10.1038/35085548.

Ellis, D.A., Martin, J.W., De Silva, A.O., Mabury, S.A., Hurley, M.D., Andersen, M.P.S., and Wallington, T.J. (2004). Degradation of fluorotelomer alcohols: A likely atmospheric source of perfluorinated carboxylic acids. *Environ. Sci. Technol.* 38 (12), 3316–3321. Doi: 10.1021/es049860w.

Emptage, M., Tabinowski, J., and Odom, J.M. (1997). Effect of fluoroacetates on methanogenesis in samples from selected methanogenic environments. *Environ. Sci. Technol.* 31(3), 732–734. Doi: 10.1021/es9603822.

Farman, J.C., Gardiner, B.G., and Shanklin, J.D. (1985). Large losses of total ozone in antarctica reveal ClOX/NOX interaction *Nature*, 315(6016), 207–210. Doi: 10.1038/315207a0.

Frank, H., Christoph, E.H., Holm-Hansen, O., and Bullister, J.L. (2002). Trifluoroacetate in ocean waters. *Environ. Sci. Technol.* 36(1), 12–15. Doi: 10.1021/es0101532.

Frank, H., Klein, A., and Renschen, D. (1996). Environmental trifluoroacetate. *Nature*, 382(6586), 34–34. Doi: 10.1038/382034a0.

Franklin, J. (1993). The atmospheric degradation and impact of 1,1,1,2-tetrafluoroethane (hydrofluorocarbon 134a). *Chemosphere* 27(8), 1565–1601. Doi: 10.1016/0045-6535(93)90251-y.

Fraser, J.M., and Kaminsky, L.S. (1988). 2,2,2-trifluoroethanol intestinal and bone marrow toxicity: The role of its metabolism to 2,2,2-trifluoroacetaldehyde and trifluoroacetic acid. *Toxicol. Appl. Pharmacol.* 94(1), 84–92. Doi: 10.1016/0041-008x(88)90339-0.

Guo, J.Y., Zhai, Z.H., Wang, L., Wang, Z.Y., Wu, J., Zhang, B.Y., and Zhang, J.B. (2017). Dynamic and thermodynamic mechanisms of TFA adsorption by particulate matter. *Environ. Pollut.* 225, 175–183. Doi: 10.1016/j.envpol.2017.03.049.

Hankins, D.C., and Kharasch, E.D. (1997). Determination of the halothane metabolites trifluoroacetic acid and bromide in plasma and urine by ion chromatography. *J. Chromatograp. B.* 692(2), 413–418. Doi: 10.1016/s0378-4347(96)00527-0.

Hanson, M.L., Sibley, P.K., Mabury, S.A., Solomon, K.R., and Muir, D.C.G. (2002). Trichloroacetic acid (TCA) and trifluoroacetic acid (TFA) mixture toxicity to the macrophytes Myriophyllum spicatum and Myriophyllum sibiricum in aquatic microcosms. *Sci. Total Environ.* 285(1–3), 247–259, Article Pii s0048-9697(01)00955-x. Doi: 10.1016/s0048-9697(01)00955-x.

Hanson, M.L., and Solomon, K.R. (2004). Haloacetic acids in the aquatic environment. Part I: macrophyte toxicity. *Environ. Pollut.* 130(3), 371–383. Doi: 10.1016/j.envpol.2003.12.016.

Harnisch, J., Frische, M., Borchers, R., Eisenhauer, A., and Jordan, A. (2000). Natural fluorinated organics in fluorite and rocks. *Geophys. Res. Lett.* 27(13), 1883–1886. Doi: 10.1029/2000gl008488.

Henne, S., Shallcross, D.E., Reimann, S., Xiao, P., Brunner, D., O'Doherty, S., and Buchmann, B. (2012). Future emissions and atmospheric fate of HFC-1234yf from mobile air conditioners in Europe. *Environ. Sci. Technol.* 46(3), 1650–1658. Doi: 10.1021/es2034608.

Holaday, D.A. (1977). Absorption, biotransformation, and storage of halothane. *Environ. Health Perspect.* 21(DEC), 165–169. Doi: 10.2307/3428517.

Hori, H., Yamamoto, A., Hayakawa, E., Taniyasu, S., Yamashita, N., and Kutsuna, S. (2005). Efficient decomposition of environmentally persistent perfluorocarboxylic acids by use of persulfate as a photochemical oxidant. *Environ. Sci. Technol.* 39(7), 2383–2388. Doi: 10.1021/es0484754.

Hurley, M.D., Andersen, M.P.S., Wallington, T.J., Ellis, D.A., Martin, J.W., and Mabury, S.A. (2004). Atmospheric chemistry of perfluorinated carboxylic acids: Reaction with OH radicals and atmospheric lifetimes. *J. Phys. Chem. A.* 108(4), 615–620. Doi: 10.1021/jp036343b.

Janda, J., Nodler, K., Brauch, H.J., Zwiener, C., and Lange, F.T. (2019). Robust trace analysis of polar (C-2-C-8) perfluorinated carboxylic acids by liquid chromatography-tandem mass spectrometry: method development and application to surface water, groundwater and drinking water. *Environ. Sci. Pollut. Res.* 26(8), 7326–7336. Doi: 10.1007/s11356-018-1731-x.

Jordan, A., and Frank, H. (1999). Trifluoroacetate in the environment. Evidence for sources other than HFC/HCFCs. *Environ. Sci. Technol.* 33(4), 522–527. Doi: 10.1021/es980674y.

Kazil, J., McKeen, S., Kim, S.W., Ahmadov, R., Grell, G.A., Talukdar, R.K., and Ravishankara, A.R. (2014). Deposition and rainwater concentrations of trifluoroacetic acid in the United States from the use of HFO-1234yf. *J. Geophys. Res.-Atmos.* 119(24), 14059–14079. Doi: 10.1002/2014jd022058.

Kim, B.R., Suidan, M.T., Wallington, T.J., and Du, X. (2000). Biodegradability of trifluoroacetic acid. *Environ. Eng. Sci.* 17(6), 337–342. Doi: 10.1089/ees.2000.17.337.

Kotamarthi, V.R., Rodriguez, J.M., Ko, M.K.W., Tromp, T.K., Sze, N.D., and Prather, M.J. (1998). Trifluoroacetic acid from degradation of HCFCs and HFCs: A three-dimensional modeling study. *J. Geophys. Res.-Atmos.* 103(D5), 5747–5758. Doi: 10.1029/97jd02988.

Kutsuna, S., and Hori, H. (2008). Experimental determination of Henry's law constants of trifluoroacetic acid at 278–298 K. *Atmos. Environ.* 42(7), 1399–1412. Doi: 10.1016/j.atmosenv.2007.11.009.

Lide, D.E. (2003). *CRC Handbook of Chemistry and Physics.* Boca Raton: CRC Press.

Lifongo, L.L., Bowden, D.J., and Brimblecombe, P. (2010). Thermal degradation of haloacetic acids in water. *Int. J. Phys. Sci.* 5(6), 738–747. <Go to ISI>://WOS: 000280354400015.

Luecken, D.J., Waterland, R.L., Papasavva, S., Taddonio, K.N., Hutzell, W.T., Rugh, J.P., and Andersen, S.O. (2010). Ozone and TFA impacts in North America from degradation of 2,3,3,3-Tetrafluoropropene (HFO-1234yf), a potential greenhouse gas replacement. *Environ. Sci. Technol.* 44(1), 343–348. Doi: 10.1021/es902481f.

Martin, J.W., Mabury, S.A., Wong, C.S., Noventa, F., Solomon, K.R., Alaee, M., and Muir, D.C.G. (2003). Airborne haloacetic acids. *Environ. Sci. Technol.* 37(13), 2889–2897. Doi: 10.1021/es026345u.

Maruthamuthu, P., and Huie, R.E. (1995). Ferric ion assisted photooxidation of haloacetates. *Chemosphere* 30(11), 2199–2207. Doi: 10.1016/0045-6535(95)00091-l.

Maruthamuthu, P., Padmaja, S., and Huie, R.E. (1995). Rate constants for some reactions of free radicals with haloacetates in aqueous solution. *Int. J. Chem. Kinet.* 27(6), 605–612. Doi: 10.1002/kin.550270610.

Milne, J.B., and Parker, T.J. (1981). Dissociation constant of aqueous trifluoroacetic acid by cryoscopy and conductivity. *J. Sol. Chem.* 10(7), 479–487. Doi: 10.1007/bf00652082.

Mogelberg, T.E., Nielsen, O.J., Sehested, J., Wallington, T.J., and Hurley, M.D. (1994). Atmospheric chemistry of CF3COOH. Kinetics of the reaction with OH radicals. *Chem. Phys. Lett.* 226(1–2), 171–177. Doi: 10.1016/0009-2614(94)00692-x.

Molina, M.J., and Rowland, F.S. (1974). Stratospheric sink for chlorofluoromethanes - chlorine atomic-catalysed destruction of ozone. *Nature*, 249(5460), 810–812. Doi: 10.1038/249810a0.

Montzka, S.A., Dutton, G.S., Yu, P.F., Ray, E., Portmann, R.W., Daniel, J.S. et al., (2018). An unexpected and persistent increase in global emissions of ozone-depleting CFC-11. *Nature*, 557(7705), 413-+. Doi: 10.1038/s41586-018-0106-2.

Namazian, M., Zakery, M., Noorbala, M.R., and Coote, M.L. (2008). Accurate calculation of the pK(a) of trifluoroacetic acid using high-level ab initio calculations. *Chem. Phys. Lett.* 451(1–3), 163–168. Doi: 10.1016/j.cplett.2007.11.088.

Nielsen, O.J., Javadi, M.S., Andersen, M.P.S., Hurley, M.D., Wallington, T.J., and Singh, R. (2007). Atmospheric chemistry of CF3CF=CH2: Kinetics and mechanisms of gas-phase reactions with Cl atoms, OH radicals, and O-3. *Chem. Phys. Lett.* 439 (1–3), 18–22. Doi: 10.1016/j.cplett.2007.03.053.

Nielsen, O.J., Scott, B.F., Spencer, C., Wallington, T.J., and Ball, J.C. (2001). Trifluoroacetic acid in ancient freshwater. *Atmos. Environ.* 35(16), 2799–2801. Doi: 10.1016/s1352-2310(01)00148-0.

Ochoa-Herrera, V., Field, J.A., Luna-Velasco, A., and Sierra-Alvarez, R. (2016). Microbial toxicity and biodegradability of perfluorooctane sulfonate (PFOS) and shorter chain perfluoroalkyl and polyfluoroalkyl substances (PFASs). *Environ. Sci.-Proc. Impacts* 18(9), 1236–1246. Doi: 10.1039/c6em00366d.

Pehkonen, S.O., Siefert, R.L., and Hoffmann, M.R. (1995). Photoreduction of iron oxyhydroxides and the photooxidation of halogenated acetic acids. *Environ. Sci. Technol.* 29(5), 1215–1222. Doi: 10.1021/es00005a012.

Qian, Y.J., Guo, X., Zhang, Y.L., Peng, Y., Sun, P.Z., Huang, C.H., Niu, J.F., Zhou, X.F., and Crittenden, J.C. (2016). Perfluorooctanoic acid degradation using UV-persulfate process: Modeling of the degradation and chlorate formation. *Environ. Sci. Technol.* 50(2), 772–781. Doi: 10.1021/acs.est.5b03715.

Richard, A.M., and Hunter, E.S. (1996). Quantitative structure-activity relationships for the developmental toxicity of haloacetic acids in mammalian whole embryo culture. *Teratology* 53(6), 352–360. Doi: 10.1002/(sici)1096-9926(199606)53:6<352::Aid-tera6>3.0.Co;2-1.

Richey, D.G., Driscoll, C.T., and Likens, G.E. (1997). Soil retention of trifluoroacetate. *Environ. Sci. Technol.* 31(6), 1723–1727. Doi: 10.1021/es960649x.

Rigby, M., Park, S., Saito, T., Western, L.M., Redington, A.L., Fang, X. et al., (2019). Increase in CFC-11 emissions from eastern China based on atmospheric observations. *Nature* 569(7757), 546-+. Doi: 10.1038/s41586-019-1193-4.

Rollins, A., Barber, J., Elliott, R., and Wood, B. (1989). Xenobiotic monitoring in plants by F-19 and H-1 nuclear magnetic resonance imaging and spectroscopy. Uptake of trifluoroacetic acid in Lycopersicon esculentum. *Plant Physiol.* 91(4), 1243–1246. Doi: 10.1104/pp.91.4.1243.

Rompp, A., Klemm, O., Fricke, W., and Frank, H. (2001). Haloacetates in fog and rain. *Environ. Sci. Technol.* 35(7), 1294–1298. Doi: 10.1021/es0012220.

Russell, M.H., Hoogeweg, G., Webster, E.M., Ellis, D.A., Waterland, R.L., and Hoke, R.A. (2012). TFA from HFO-1234yf: Accumulation and aquatic risk in terminal water bodies. *Environ. Toxicol. Chem.* 31(9), 1957–1965. Doi: 10.1002/etc.1925.

Scheurer, M., Nodler, K., Freeling, F., Janda, J., Happel, O., Riegel, M., Muller, U., Storck, F.R., Fleig, M., Lange, F.T., Brunsch, A., and Brauch, H.J. (2017). Small, mobile, persistent: Trifluoroacetate in the water cycle - Overlooked sources, pathways, and consequences for drinking water supply. *Water Res.* 126, 460–471. Doi: 10.1016/j.watres.2017.09.045.

Schwarzbach, S.E. (1995). Ozone depletion - CFC alternatives under a cloud. *Nature* 376(6538), 297–298. Doi: 10.1038/376297a0.

Scott, B.F., Macdonald, R.W., Kannan, K., Fisk, A., Witter, A., Yamashita, N., Durham, L., Spencer, C., and Muir, D.C.G. (2005). Trifluoroacetate profiles in the Arctic, Atlantic, and Pacific Oceans. *Environ. Sci. Technol.* 39(17), 6555–6560. Doi: 10.1021/es047975u.

Scott, B.F., Mactavish, D., Spencer, C., Strachan, W.M.J., and Muir, D.C.G. (2000). Haloacetic acids in Canadian lake waters and precipitation. *Environ. Sci. Technol.* 34(20), 4266–4272. Doi: 10.1021/es9908523.

Scott, B.F., Moody, C.A., Spencer, C., Small, J.M., Muir, D.C.G., and Mabury, S.A. (2006). Analysis for perfluorocarboxylic acids/anions in surface waters and precipitation using GC-MS and analysis of PFOA from large-volume samples. *Environ. Sci. Technol.* 40(20), 6405–6410. Doi: 10.1021/es061131o.

Scott, B.F., Spencer, C., Mabury, S.A., and Muir, D.C.G. (2006). Poly and perfluorinated carboxylates in north American precipitation. *Environ. Sci. Technol.* 40(23), 7167–7174. Doi: 10.1021/es061403n.

Scott, B.F., Spencer, C., Martin, J.W., Barra, R., Bootsma, H.A., Jones, K.C., Johnston, A.E., and Muir, D.C.G. (2005). Comparison of haloacetic acids in the environment of the northern and southern hemispheres. *Environ. Sci. Technol.* 39(22), 8664–8670. Doi: 10.1021/es050118l.

Scott, B.F., Spencer, C., Marvin, C.H., MacTavish, D.C., and Muir, D.C.G. (2002). Distribution of haloacetic acids in the water columns of the Laurentian Great Lakes and Lake Malawi. *Environ. Sci. Technol.* 36(9), 1893–1898. Doi: 10.1021/es011156h.

Singh, R.K., Fernando, S., Baygi, S.F., Multari, N., Thagard, S.M., and Holsen, T.M. (2019). Breakdown products from Perfluorinated Alkyl Substances (PFAS) degradation in a plasma-based water treatment process. *Environ. Sci. Technol.* 53(5), 2731–2738. Doi: 10.1021/acs.est.8b07031.

Smit, M.F., van Heerden, P.D.R., Pienaar, J.J., Weissflog, L., Strasser, R.J., and Kruger, G.H.J. (2009). Effect of trifluoroacetate, a persistent degradation product of fluorinated hydrocarbons, on Phaseolus vulgaris and Zea mays. *Plant Physiol. Biochem.* 47(7), 623–634. Doi: 10.1016/j.plaphy.2009.02.003.

Solomon, K.R., Velders, G.J.M., Wilson, S.R., Madronich, S., Longstreth, J., Aucamp, P.J., and Bornman, J.F. (2016). Sources, fates, toxicity, and risks of trifluoroacetic acid and its salts: Relevance to substances regulated under the Montreal and Kyoto Protocols. *J. Toxicol. Environ. Health-Part B-Critical Rev.* 19(7), 289–304. Doi: 10.1080/10937404.2016.1175981.

Taniyasu, S., Kannan, K., Yeung, L.W.Y., Kwok, K.Y., Lam, P.K.S., and Yamashita, N. (2008). Analysis of trifluoroacetic acid and other short-chain perfluorinated acids (C2-C4) in precipitation by liquid chromatography-tandem mass spectrometry: Comparison to patterns of long-chain perfluorinated acids (C5-C18). *Analytica Chimica Acta* 619(2), 221–230. Doi: 10.1016/j.aca.2008.04.064.

Tian, Y., Yao, Y.M., Chang, S., Zhao, Z., Zhao, Y.Y., Yuan, X.J., Wu, F.C., and Sun, H.W. (2018). Occurrence and phase distribution of neutral and ionizable per- and polyfluoroalkyl substances (PFASs) in the atmosphere and plant leaves around landfills: A case study in Tianjin, China. *Environ. Sci. Technol.* 52(3), 1301–1310. Doi: 10.1021/acs.est.7b05385.

Tromp, T.K., Ko, M.K.W., Rodriguez, J.M., and Sze, N.D. (1995). Potential accumulation of a CFC-replacement degradation product in seasonal wetlands. *Nature* 376 (6538), 327–330. Doi: 10.1038/376327a0.

Visscher, P.T., Culbertson, C.W., and Oremland, R.S. (1994). Degradation of trifluoroacetate in oxic and anoxic sediments. *Nature* 369(6483), 729–731. Doi: 10.1038/369729a0.

Vollmer, M.K., Rhee, T.S., Rigby, M., Hofstetter, D., Hill, M., Schoenenberger, F., and Reimann, S. (2015). Modern inhalation anesthetics: Potent greenhouse gases in the global atmosphere. *Geophys. Res. Lett.* 42(5), 1606–1611. Doi: 10.1002/2014gl062785.

Von Sydow, L.M., Grimvall, A.B., Boren, H.B., Laniewski, K., and Nielsen, A.T. (2000). Natural background levels of trifluoroacetate in rain and snow. *Environ. Sci. Technol.* 34(15), 3115–3118. Doi: 10.1021/es9913683.

Wallington, T.J., Hurley, M.D., Fedotov, V., Morrell, C., and Hancock, G. (2002). Atmospheric chemistry of CF3CH2OCHF2 and CF3CHClOCHF2: Kinetics and mechanisms of reaction with Cl atoms and OH radicals and atmospheric fate of CF3C(O•)HOCHF2 and CF3C(O•)ClOCHF2 radicals. *J. Phys. Chem.* A 106(36), 8391–8398. Doi: 10.1021/jp020017z.

Wallington, T.J., Schneider, W.F., Worsnop, D.R., Nielsen, O.J., Sehested, J., Debruyn, W.J., and Shorter, J.A. (1994). The environmental-impact of CFC replacements – HFCs and HCFCs. *Environ. Sci. Technol.* 28(7), A320–A326. Doi: 10.1021/es00056a002.

Wang, Q.Y., Wang, X.M., and Ding, X. (2014). Rainwater trifluoroacetic acid (TFA) in Guangzhou, South China: Levels, wet deposition fluxes and source implication. *Sci. Total Environ.* 468, 272–279. Doi: 10.1016/j.scitotenv.2013.08.055.

Wang, Z.Y., Wang, Y.H., Li, J.F., Henne, S., Zhang, B.Y., Hu, J.X., and Zhang, J.B. (2018). Impacts of the degradation of 2,3,3,3-tetrafluoropropene into trifluoroacetic acid from its application in automobile air conditioners in China, the United States, and Europe. *Environ. Sci. Technol.* 52(5), 2819–2826. Doi: 10.1021/acs.est.7b05960.

Waskell, L. (1978). Study of mutagenicity of anesthetics and their metabolites. *Mutation Res.* 57(2), 141–153. Doi: 10.1016/0027-5107(78)90261-0.

Wu, J., Martin, J.W., Zhai, Z.H., Lu, K.D., Li, L., Fang, X.K., Jin, H.B., Hu, J.X., and Zhang, J.B. (2014). Airborne trifluoroacetic acid and its fraction from the degradation of HFC-134a in Beijing, China. *Environ. Sci. Technol.* 48(7), 3675–3681. Doi: 10.1021/es4050264.

Wujcik, C.E., Cahill, T.M., and Seiber, J.N. (1999). Determination of trifluoroacetic acid in 1996–1997 precipitation and surface waters in California and Nevada. *Environ. Sci. Technol.* 33(10), 1747–1751. Doi: 10.1021/es980697c.

Wujcik, C.E., Zehavi, D., and Seiber, J.N. (1998). Trifluoroacetic acid levels in 1994–1996 fog, rain, snow and surface waters from California and Nevada. *Chemosphere* 36(6), 1233–1245. Doi: 10.1016/s0045-6535(97)10044-3.

Xie, G.Y., Cui, J.N., Zhai, Z.H., and Zhang, J.B. (2020). Distribution characteristics of trifluoroacetic acid in the environments surrounding fluorochemical production plants in Jinan, China. *Environ. Sci. Pollut. Res.* 27(1), 983–991. Doi: 10.1007/s11356-019-06689-4.

Zehavi, D., and Seiber, J.N. (1996). An analytical method for trifluoroacetic acid in water and air samples using headspace gas chromatographic determination of the methyl ester. *Analy. Chem.* 68(19), 3450–3459. Doi: 10.1021/ac960128s.

Zhai, Z.H., Wu, J., Hu, X., Li, L., Guo, J.Y., Zhang, B.Y., Hu, J.X., and Zhang, J.B. (2015). A 17-fold increase of trifluoroacetic acid in landscape waters of Beijing, China during the last decade. *Chemosphere* 129, 110–117. Doi: 10.1016/j.chemosphere.2014.09.033.

Zhang, B.Y., Zhai, Z.H., and Zhang, J.B. (2018). Distribution of trifluoroacetic acid in gas and particulate phases in Beijing from 2013 to 2016. *Sci. Total Environ.* 634, 471–477. Doi: 10.1016/j.scitotenv.2018.03.384.

Zhang, J.B., Zhang, Y., Li, J.L., Hu, J.X., Ye, P., and Zeng, Z. (2005). Monitoring of trifluoroacetic acid concentration in environmental waters in China. *Water Res.* 39(7), 1331–1339. Doi: 10.1016/j.watres.2004.12.043.

# 10

## Drift

### 10.1 Introduction

Pesticide residues and other contaminants can move downwind of target application sites and undergo transport to nearby as well as distant places. This opens pathways for exposure to nontarget organisms including people (Kurtz, 1990). Airborne contaminants can be transformed in the air by a number of processes including oxidation and photochemical reactions; in some cases, the reactions yield products that are more toxic than the parent (Seiber et al., 1980). When the transformation product is more toxic than the parent, it is termed or called "activation." If the products are less toxic than the parent, it is termed "deactivation." Figure 10.1 illustrates the processes involved in pesticide drift.

Transect studies are useful to follow the movement and distribution of pesticides and other contaminants in the environment. These residues are carried by air and distributed downwind where they are deposited to other more remote environmental compartments such as soil, water, plant surfaces, and biota (Aston and Seiber, 1996, 1997; Ross et al., 1990; Seiber et al., 1989; Woodrow et al., 1990). Transect studies can also be used as a technique to locate the source of a contaminant as well as its path. An example is for chlorofluorocarbons (CFCs), where transect studies identified urban areas and transportation corridors as likely sources (Frank et al., 1996).

"Flux studies" measure the amount of pesticides emitted to air, which occurs during spraying, and due to evaporation and erosion post application. Near source flux studies conducted shortly after pesticide application provide emission rates that can be used as source terms for modeling drift to nontarget sites. To comply with the California Toxic Air Contaminant Act, these "flux studies" have been conducted by or contracted out by the California Air Resources Board and Department of Pesticide Regulation for pesticides used throughout the state. Such flux and drift studies have been conducted for methyl bromide, emitted to the air after fumigation (see Chapter 8 for a detailed example for methyl bromide applied to soil); parathion and chlorpyrifos, emitted from treated orchards; and toxaphene use in cotton fields (see Chapter 3) (Seiber and Woodrow, 1981; Woodrow et al., 1983, 1997, 2001).

DOI: 10.1201/9781003217602-10

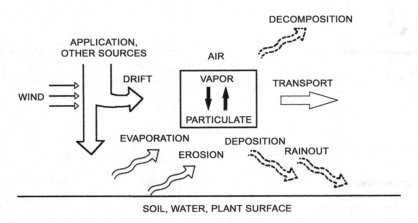

**FIGURE 10.1**
The central role of air in the transport of pesticides through the environment. (Reprinted (adapted) with permission from Seiber, J.N., Ferreira, G.A., Hermann, B., and Woodrow, J.E. (1980). Analysis of pesticidal residues in the air near agricultural treatment sites. In J. Harvey, Jr., and G. Zweig, (Eds.), *American Chemical Society Symposium Series No. 136 Pesticide Analytical Methodology*, ACS Publications. Copyright 1980 American Chemical Society.) and (Reprinted by permission from Springer Nature Woodrow, J.E., Gibson, K.A., and Seiber, J.N. (2018). Pesticides and related toxicants in the atmosphere. In P. de Voogt, (Ed.), *Reviews of Environmental Contamination and Toxicology* (Vol. 247, pp. 147–196). Cham: Springer International Publishing. https://doi.org/10.1007/398_2018_19. Copyright Springer Nature Switzerland AG 2018.)

Damage to almond and prune orchards was traced back to airborne drift of 2-methyl-4-chlorophenoxyacetic acid (MCPA) and propanil that was applied to rice (Barry and Walters, 2002; Crosby et al., 1981).

In transect studies, airborne residues are sampled downwind of sources at different times and different distances to track the downwind movement of chemicals from sources. With the aid of modeling, drift studies can estimate exposures experienced by downwind populations and allow for risk assessment to ensure compliance with the California Toxic Air Contaminant Act or other regulations.

The behavior of different chemicals provides interesting contrasts in their environmental distribution (Table 10.1). The observed drift is a direct result of both their physicochemical properties and their use patterns. Parathion and chlorpyrifos gradients decreased in concentration moving from the west, where they were applied in the Central Valley to the east in the Sierra Nevada Mountains. In kind, concentrations in the air were much higher at lower elevations. The partitioning of parathion and its activated oxidation product paraoxon in fog made red-tailed hawks and other wildlife that frequented orchards particularly vulnerable as it deposited from the fog to their feathers during wintertime dormant spraying of tree crops in areas which are characteristically foggy (Fry et al., 1998). Parathion was banned by EPA in about 1980.

**TABLE 10.1**

How Pesticide Properties and Use Impacts Distribution

|  | Chlorpyrifos | Parathion | PCB | TFA |
|---|---|---|---|---|
| Sources | Agricultural | Agricultural | Variety of sources and broad usage | HCFC breakdown, potential natural sources |
| Polarity | Moderate | Moderate | Low | High |
| Volatility | Semivolatile | Semivolatile | Semivolatile | Nonvolatile |
| Water solubility (max in surface water) | Moderate (200 ng/L) | Moderate | Low (1 ng/L) | High (10,000 ng/L, vernal pools) |
| Stability | Moderate | Moderate | Persistent | Persistent |
| Compartments | Precipitation, surface water, air, plant surfaces | Precipitation, surface water, air, plant surfaces | Biota (fish), precipitation, fog, surface water | Precipitation, air, evapoconcentration in terminal waters, water bodies |
| Concentration gradients | North to South East to West | North to South East to West | No clear gradients | North to South East to West |

*Source:* Adapted from Woodrow et al. (2018).

Chlorpyrifos, recently banned, is a semivolatile pesticide that was in widespread use on agricultural commodities (Lu, 2018). Chlorpyrifos gradients increase in use areas, and from west in the agricultural sectors to east (in the Sierra Nevada). This semivolatile and moderately persistent organophosphate (OP) pesticide partitions to fog at higher concentrations than would be predicted by its Henry's constant (see Chapter 7) and is found in rain, snow, fogwater, and other surface waters near the agricultural sources (McConnell et al., 1998).

Polychlorinated biphenyls (PCBs) are categorized as persistent organic chemicals (POPs). Until they were banned in 1979, they were used in many products in addition to pesticides, including consumer products such as paper, adhesives, and electronics. PCBs show no clear gradients due to the variety of sources and areas in which they were used. These stable and semivolatile compounds have significant airborne transport to higher elevations where they can be deposited by wind and precipitation events. Low polarity makes them fat soluble, giving them a high bioconcentration factor (BCF), as apparent from their higher concentrations at the higher trophic levels of the organisms studied (see Chapter 3).

Trifluoroacetic acid (TFA) (see Chapter 9) is a stable breakdown product of hydrochlorofluorocarbons (HCFCs), replacements of CFCs. TFA is nonvolatile, polar, and very water soluble. In vernal pools and other seasonal wetlands, concentrations up to 10,000 ng/L have been detected. These concentrations are high in these terminal water bodies due to evapoconcentration. It is thus a good tracer for contaminants to identify their sources (Cahill and Seiber, 2000; Cahill et al., 1999; Frank et al., 1996). Other stable and polar chemical breakdown products include dimethyl phosphate and diethyl phosphate from OP pesticides. These chemicals might offer targets for following transport and fate from high OP use areas similar to TFA from Freon replacement breakdown.

## 10.2 Drift to Sensitive Ecosystems: From California's Central Valley to the Sierra Nevada

Agriculture in California's Central Valley is a source of pesticides that are transported in the air to sensitive ecosystems such as in the Sierra Nevada Mountain Range (Woodrow et al., 2018). Air moves eastward from the Pacific Ocean to inland areas (Figure 10.2). Along this route, the airmass accumulates contaminants from industrialized coastal, vehicle-laden areas such as the San Francisco Bay area. Northerly winds from the coast transport dust and vapor containing pesticides from the northern Sacramento valley, and southerly winds bring the airmass to agricultural San Joaquin Valley which

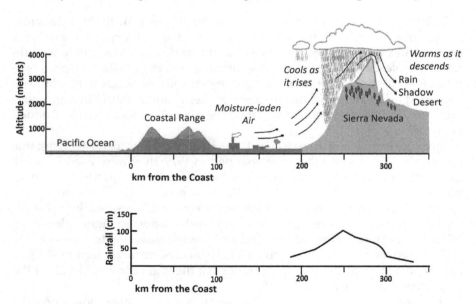

**FIGURE 10.2**
Cross section of California (Bradford et al., 2010; Hayes, 1984) (Reprinted by permission from Springer Nature Woodrow, J.E., Gibson, K.A., and Seiber, J.N. (2018). Pesticides and related toxicants in the atmosphere. In P. de Voogt, (Ed.), *Reviews of Environmental Contamination and Toxicology* (Vol. 247, pp. 147–196). Cham: Springer International Publishing. https://doi.org/10.1007/398_2018_19. Copyright Springer Nature Switzerland AG 2018.)

consists of a network of canals, streams, flooded and submerged or irrigated fields. Pesticides from the agricultural regions of the Central Valley volatilize and drift from warm parts of the valley to cooler high elevation regions of the Sierra Nevada Mountains. Eventually the airmass moves to the foothill of the Sierra's and then to higher elevations of the western slopes of the Sierra's.

During transport, pesticides undergo degradation and are scrubbed from the air by wet/dry deposition. Wet deposition via snow and rain is prevalent during winter and spring. The final destinations of deposited pesticides and/or pesticide degradation products are to surface waters, soil (and eventually groundwater), plant surfaces, and biota.

In one study, high volume air samples were collected, in addition to dry deposition and surface waters at various points in the valley and Sequoia National Park (118—3,322 m elevation) (LeNoir et al., 1999). Surface waters contained residues of trifluralin, diazinon, chlorothalonil, chlorpyrifos, chlorpyrifos oxon, malathion, and endosulfan I/II. Those most heavily used in summertime: trifluralin, chlorpyrifos, and endosulfan, were the most abundant in surface waters. Air samples additionally contained endosulfan sulfate; whereas only chlorothalonil, chlorpyrifos, chlorpyrifos oxon, and endosulfan I/II were detected in dry deposition samples. Tested fish species (trout and stonefly) had concentrations well below 96-hour $LC_{50}$ values, but only the more sensitive amphipods would likely suffer from the effects at these levels.

A series of studies have confirmed the long-range transport of PCBs that contaminate water bodies, as opposed to coming from local sources. A transect study of the lakes and streams of the Tahoe Basin and Lake Tahoe itself detected PCBs in both water and biota (trout) with similar profiles (Datta et al., 1998a). In a companion study, PCB congener profiles were compared between Lake Tahoe and Marlette Lake and were found to be the same. Since Marlette Lake is isolated and off-limits to vehicles and recreational fishing, and monitored by the Nevada Division of Wildlife, detected PCBs were attributed to long-distance transport as opposed to being generated from local sources (Datta et al., 1998b).

Decline of sensitive amphibian populations in the Sierra Nevada has been attributed to drift of PCBs and chlorinated pesticides such as dichlorodiphenyltrichloroethane (DDT) and toxaphene, which may act as endocrine disrupters (National Pesticide Information Center, 2020). Earlier studies sampled frogs from the Sierra Nevada Mountains and found measurable dichlorodiphenyldichloroethylene (DDE) which provided evidence of DDT contamination from prior years (Cory et al., 1970). DDT was banned in 1972 but its residues persist, especially its long-lived breakdown product, DDE. A concentration gradient was noted: frogs in the western slopes of the mountains had higher concentrations of DDT/DDE, indicating they originated from California's central valley. However, in the case of toxaphene, which was used heavily in the cotton fields of California before 1975, the low levels detected in pacific tree frogs downwind of the source did not provide convincing evidence linking declining populations to toxaphene drift (Angermann et al., 2002).

In a later study, air, sediment, and tadpoles were collected from 28 sites in Sequoia and Kings Canyon National Parks. Residues frequently found in sediments and tadpoles included current and historic use pesticides, PCBs, and others. Concentration of all chemicals was very low on the order of 10 pg/ $m^3$ (Bradford et al., 2010). These studies showed clearly that pesticides from California's Central Valley are deposited in the Sierra Nevada Mountains.

Drift studies indicated that pesticides applied in the southern Central Valley of California (agricultural) are transported through the air to the Sequoia National Park in the Sierra Nevada Mountains (rural) (Aston and Seiber, 1996, 1997). It was estimated that up to 16 kg of chlorpyrifos and its oxon are deposited onto plant foliage over the course of a summer. These studies revealed that the pines that dominate these forests are a major sink for pesticides applied during summertime. Pine needles and other foliage work in a similar manner to solid-phase microextraction devices (see Chapter 6). They are an important endpoint to study due to the relevance to sensitive wildlife that live in trees.

Yellow-legged frogs (*Rana muscosa*) in the Tablelands region of the Sequoia National Forest were thought to be in decline due to the direct winds from the Central Valley of California. After a failed attempt to reintroduce yellow-legged frogs, pesticide levels in water and surviving Tablelands frogs were measured and compared to samples from the Sixty Lakes Basin in Kings Canyon National Park which had a healthy frog population (Fellers et al., 2004). γ-Chlordane and trans-nonachlor were detected in higher levels in Tablelands frogs than in frogs from the Sixty Lakes Basin. OP insecticides, chlorpyrifos and diazinon, were found in higher levels in water from the Tablelands than in the Sixty Lakes Basin.

Transect studies of wetlands spanning from the Pacific coast to the Cascades or Sierra Nevada Mountains, cutting through Lassen Volcanic National Park, Lake Tahoe, Yosemite National Park, and Sequoia National Park, revealed the presence of 23 pesticides in water, sediment, and biota (in this case, Pacific chorus frogs) (Fellers et al., 2013). Individual pesticides did not exceed the known lethal or sublethal concentrations, however, up to six different pesticides, including trifluralin, α-endosulfan, chlordanes, and trans-nonachlor, were found in adult frog tissues. Chlorpyrifos oxon in sediment and total endosulfans exceeded safe levels for frogs. β-Endosulfan, banned in 2011 by the Stockholm convention due to its neurotoxic effects to humans and animals, was present in almost all samples. Lower cholinesterase activities were measured in areas with the highest population decline. Chlorpyrifos has been proven to reduce cholinesterase activities in tadpoles by up to 43% (Widder and Bidwell, 2006). Interestingly, it does not affect tadpole survival because of pesticide-induced predator mortality.

More than 90 current use pesticides and their degradates were measured in a transect study at seven different elevations in the mountains of Northern California, downwind from high pesticide use areas of the Central Valley (Smalling et al., 2013). The most frequently detected were two fungicides, pyraclostrobin and tebuconazole, and the herbicide, simazine. The data indicated that amphibian populations in remote counties are accumulating pesticides, and tissue concentrations correlated with pesticides used upwind. The evidence shows that the frogs may be more reliable indicators of exposure that water or sediment.

The mountain ranges in the western United States are ideal for study given they are located near both agricultural and urban sources of toxics and have fragile ecosystems with sensitive wildlife. Our work and that of others have clearly shown that pesticides are transported through air to remote locations, and these residues are accumulated by sensitive biota such as frogs. The urgency of current and future work is to connect these residues with health. Studies need to focus on the current use and persistent pesticides to assess biological impact. Bioassays, including cholinesterase inhibition (e.g., related to chlorpyrifos and parathion exposures), and endocrine disruption (e.g., related to chlordane and other chlorinated residues, possibly trifluralin exposures), should be routinely analyzed. Coupling transect studies to mesocosms and *in vitro* testing that measures tissue levels of pesticides under different waterborne exposures would enable correlation with what is measured in transect studies, particularly in areas where wildlife are experiencing adverse effects. Additional targets may be found in fragile ecosystem for birds, fish, bees, and other wildlife.

Chlorpyrifos is more toxic to aquatic organisms than diazinon (amphipods $LC_{50}$: 110 vs 200 ng/mL), and moving up the food chain, organisms are more tolerant ($LC_{50}$ for chlorpyrifos in fathead minnow: 0.1 vs in lake trout 98 ng/ mL), however, chlorpyrifos bioaccumulates (Eisler, 1986; Odenkirchen and Eisler, 1988). If measured residues in water and sediment are too low to cause toxicity, we need to look higher in the food chain such as lake trout. Such an undertaking will require the close coordination of public agencies that study water sources, recreational areas, and wildlife habitat, an example of such study is the Sierra Nevada–Southern Cascades study that includes air monitoring and other research projects still in progress. This may address questions such as those raised in this and other chapters.

## 10.3 Pesticide Drift to Nontarget Crops

Pesticide drift through air to nearby or downwind areas can cause damage to nontarget crops sensitive to the chemical. This can include visible damage (scorched, burnt, dead) vegetation, such as occurs in some uses of contact herbicides, for example, herbicides like the phenoxy class (MCPA, 2,4-D) bipyridyls (paraquat), glyphosate, dicamba, and others.

In the case of systemic herbicides like glyphosate, drift to nontarget crops can cause death to the crops (annuals and/or perennials) and thus incur substantial economic damage, for example, the need to replant orchards and vineyards. Training of farmers and PCOs backed by regulatory guidance on the product label and by the agricultural commissioner, cooperative extension, and other experienced personnel can help minimize drift damage.

An associated problem occurs with drift of inadvertent residues via fogwater (see Chapter 7) to a crop on which the chemical is not registered for use. This can entail an illegal residue finding, resulting in destruction of the crop or preventing its entry into trade channels. If it is due to long range drift such that it is not clear where the residue originated, and it is determined that the residues are not a threat to humans or wildlife, regulatory relief can sometimes be obtained by adding the unintended crop that received the drift residue to the product label so that it becomes then a legal residue at harvest, with appropriate tolerance.

An unexpected source of volatilization with drift occurred following application of systemic carbamate insecticides to paddy rice at the International Rice Research Institute (IRRI) in the Philippines. Application was by broadcast of a granular formulation of carbofuran with the expectation that granules would release the active ingredient such that it would be taken up by roots of rice plants, then translocate through stems to leaves, controlling brown planthoppers and other pests in the process. Air sampling above the crop canopy showed that much of the parent carbamate plus some plant metabolites had volatilized. Siddaramappa et al. (1978) showed that the internal residue exited from plants via guttation fluid (see Figure 10.3), leaving an

**FIGURE 10.3**
This photo shows a chamber for studying uptake of a systemic insecticide, carbofuran, by rice seedlings. The carbofuran, administered in the rootzone, is translocated through stems to leaves, where guttation allows it to exit from the leaf interior to outside surface. Once on the outside surface, it can volatilize to air and be distributed by air currents elsewhere, and or be a source of exposure for farmworkers transplanting rice seedlings into production paddies. Volatilization occurs for pesticides applied to plant surfaces, but also in many other circumstances, like this one where residues are taken up by plants from soil or water. Soil or plant metabolism can co-occur, so that metabolites can volatilize along with the parent pesticide, complicating the volatilization of individual pesticides. Whether this occurs for systemic herbicides like phenoxy acids, glyphosate, and others warrants study.

external residue which subsequently volatized. The carbamate carbofuran did not leave a residue in grain at harvest, so the rice could be safely consumed. Tilapia fish reared in the same paddy did not show ill effects from carbamates and were free of carbamate residue at harvest. Thus, consumers benefited from increased yields of rice along with a protein (fish) source for their diets from the same paddy.

## 10.4 Dicamba Drift, an Ongoing Debate

A current as yet unresolved issue involves alleged vapor drift of dicamba herbicide from soybeans to nontarget crops located some distance from the initial soybean-treated area.

Dicamba (3,6-dichloro-2-methoxy benzoic acid) is a post-emergent, growth regulator herbicide used in agriculture for control of broad leaf weeds. The herbicide is volatile (4.5 mPa, 25°C). So, it must be applied as a salt, which, by definition, is normally nonvolatile. Drift to nontarget fields during application can be minimized by using ground rigs with height-adjusted (~24 in.) spray nozzles that produce aerosols of sufficient size that quickly deposit onto the target plants under optimum environmental conditions of temperature (<30°C) and wind speed (<10 mph). However, in practice, regardless of the dicamba formulation, damage to nearby, nontarget fields continues, implying that the herbicide volatilizes post application from treated fields, sometimes called "vapor drift." This was observed early on for the dimethylamine salt and sodium salt formulations and continues to be observed for the newer diglycolamine formulations. A number of field studies have identified factors that contribute to post-application volatilization: (1) temperature (Behrens and Lueschen, 1979; Mueller and Steckel, 2019), (2) pH of the tank mix (Mueller and Steckel, 2019), (3) pH of the soil surface (Oseland et al., 2020), and (4) leaf surface in the area treated (Behrens and Lueschen, 1979). Only the factors that had a direct bearing on the stability of the dicamba salt are included here. Other factors that have been associated with volatilization and nontarget damage include the following (Behrens and Lueschen, 1979; Bish et al., 2019): (1) application rate; (2) application time of day; (3) effect of dicamba formulation; (4) rainfall and relative humidity; (5) time after application; and (6) wind speed and atmospheric stability.

Obviously, it's not the salt that is volatilizing. A strong contender is the dicamba acid. Support for this comes from studies that have measured dicamba in air (ng and ng/m$^3$) post application (Bish et al., 2019; Mueller and Steckel, 2019). Acid pH will cause a dicamba salt to dissociate. For example, mixing Roundup PowerMax® with Xtendimax® VaporGrip® Technology (diglycolamine salt) will lower the pH of the tank mix, leading to greater dicamba volatilization; also, an acid pH soil surface will lead to greater volatilization. What is not known is the stability of the diglycolamine salt of dicamba. The amine salt can be viewed as a conjugate acid. So, the question is: what is the pKa of the salt? For example, the pKa for the conjugate acid of ethanolamine (and the amine itself) is 9.5, characteristic of a strong base. The diglycolamine salt of dicamba is probably of a similar order of magnitude, implying that the salt should be very stable (pKa data are not available). However, at acid pH (tank mix/soil), protonated diglycolamine may be more stable than the salt, depending on the pKa of its conjugate acid. Perhaps, this is where the effect of the temperature factor also comes into play, in addition to promoting increased volatilization of the already formed dicamba acid; volatility increases up to about 30°C. Another factor that might affect the stability of the salt is treated leaf surface area. Apparently, as the water-based tank mix dries on the leaf surfaces, the dicamba salt, as it interacts with the leaf cuticle, becomes less stable, leading to volatile dicamba. For larger treated surface areas, more of the volatile component is available to damage nontarget, susceptible crops. An additional consideration is the exposure of target leaf residues to sunlight. However, partly due to a low quantum yield (<0.02), photolysis is a relatively minor dissipation process compared to the timescale of dicamba volatilization and atmospheric dispersion (Aguer et al., 2000; Waite et al., 2005).

## 10.5 Conclusions

Lessons learned from chemical residue behavior in the air can be applied in responding to airborne exposures to other toxicants including metals such as mercury (that volatilizes as methyl mercury from the ocean particularly in near-shore ocean spray and surf) and cadmium, allergens, and infective viruses. Modeling frameworks developed for pesticide monitoring could also be used to address emerging concerns (that thus far have sparse proof) of transport and deposition of pathogens such as *Salmonella* and *E. coli* 0157H7 to nearby almonds or leafy greens, thought to travel through air from such sources as nearby feedlots for dairy cows and other farm animals.

The issue of exposure of people to pesticides in air or those deposited has been raised in several communities where pesticides are used in or around orchards or in crop land. Of particular interest are in ongoing community, agency, and grower discussions in Lompoc, McFarland, and other communities in California, Oregon, and Washington. In some cases, county or municipalities have enacted ordinances to ban or restrict application of conventional/synthetic pesticides, leaving only the options of using safer chemicals like biopesticides or alternative pest management strategies (see Chapter 12 for more on biopesticides).

# References

Aguer, J.-P., Blachère, F., Boule, P., Garaudee, S., and Guillard, C. (2000). Photolysis of dicamba (3, 6-dichloro-2-methoxybenzoic acid) in aqueous solution and dispersed on solid supports. *Int. J. Photoenergy* 2, 81–86.

Angermann, J.E., Fellers, G.M., and Matsumura, F. (2002). Polychlorinated biphenyls and toxaphene in Pacific tree frog tadpoles (Hyla regilla) from the California Sierra Nevada, USA. *Environ. Toxicol. Chem.* 21, 2209–2215.

Aston, L., and Seiber, J. (1996). Exchange of airborne organophosphorus pesticides with pine needles. *J. Environ. Sci. Health Part B.* 31, 671–698.

Aston, L.S., and Seiber, J.N. (1997). Fate of summertime airborne organophosphate pesticide residues in the Sierra Nevada mountains. *J. Environ. Qual.* 26, 1483–1492.

Barry, T., and Walters, J. (2002). *Characterization of Propanil Prune Foliage Residues as Related to Propanil Use Patterns in the Sacramento Valley.* State of California, Environmental Protection Agency, Department of Pesticide Regulation, Environmental Monitoring Branch.

Behrens, R., and Lueschen, W. (1979). Dicamba volatility. *Weed Sci.* 486–493.

Bish, M.D., Farrell, S.T., Lerch, R.N., and Bradley, K.W. (2019). Dicamba losses to air after applications to soybean under stable and nonstable atmospheric conditions. *J. Environ. Qual.* 48, 1675–1682.

Bradford, D.F., Stanley, K., McConnell, L.L., Tallent-Halsell, N.G., Nash, M.S., and Simonich, S.M. (2010). Spatial patterns of atmospherically deposited organic contaminants at high elevation in the southern Sierra Nevada mountains, California, USA. *Environ. Toxicol. Chem.* 29, 1056–1066.

Cahill, T.M., Benesch, J.A., Gustin, M.S., Zimmerman, E.J., and Seiber, J.N. (1999). Simplified method for trace analysis of trifluoroacetic acid in plant, soil, and water samples using headspace gas chromatography. *Anal. Chem.* 71, 4465–4471.

Cahill, T.M., and Seiber, J.N. (2000). Regional distribution of trifluoroacetate in surface waters downwind of urban areas in Northern California, USA. *Environ. Sci. Technol.* 34, 2909–2912.

Cory, L., Fjeld, P., and Serat, W. (1970). Distribution patterns of DDT residues in the Sierra Nevada mountains. *Pestic. Monit. J.* 3, 204–211.

Crosby, D.G., Li, M.-Y., Seiber, J.N., and Winterlin, W.L. (1981). *Environmental Monitoring of MCPA in relation to Orchard Contamination.* Davis, California: California Department of Food and Agriculture.

Datta, S., Hansen, L., McConnell, L., Baker, J., LeNoir, J., and Seiber, J. (1998a). Pesticides and PCB contaminants in fish and tadpoles from the Kaweah River Basin, California. *Bull. Environ. Contam. Toxicol.* 60, 829–836.

Datta, S., McConnell, L.L., Baker, J.E., LeNoir, J., and Seiber, J.N. (1998b). Evidence for atmospheric transport and deposition of polychlorinated biphenyls to the Lake Tahoe Basin, California– Nevada. *Environ. Sci. Technol.* 32, 1378–1385.

Eisler, R. (1986). *Diazinon Hazards to Fish, Wildlife, and Invertebrates: A Synoptic Review.* Laurel, MD: Fish and Wildlife Service.

Fellers, G.M., McConnell, L.L., Pratt, D., and Datta, S. (2004). Pesticides in mountain yellow-legged frogs (Rana muscosa) from the Sierra Nevada Mountains of California, USA. *Environ. Toxicol. Chem. Int. J.* 23, 2170–2177.

Fellers, G.M., Sparling, D.W., McConnell, L.L., Kleeman, P.M., and Drakeford, L. (2013). Pesticides in amphibian habitats of central and northern California, USA. In *Occurrence, Fate and Impact of Atmospheric Pollutants on Environmental and Human Health*. (pp. 123–150). Washington, D.C.: ACS Publications.

Frank, H., Klein, A., and Renschen, D. (1996). Environmental trifluoroacetate. *Nature* 382, 34–34.

Fry, D., Wilson, B., Ottum, N., Yamamoto, J., Stein, R., Seiber, J., McChesney, M., and Richardson, E. (1998). Radiotelemetry and GIS computer modeling as tools for analysis of exposure to organophosphate pesticides in Red-tailed Hawks. *Radiotelem. Appl. Wildl. Toxicol. Stud. Soc. Environ. Toxicol. Chem. Pensacola FL* 67–83.

Hayes, T.P. (1984). *California Surface Wind Climatology*. Irvine, CA: California Air Resources Board.

Kurtz, D.A. (1990). *Long Range Transport of Pesticides*. Chelsea, Michigan: Lewis Publishers, Inc..

LeNoir, J.S., McConnell, L.L., Fellers, G.M., Cahill, T.M., and Seiber, J.N. (1999). Summertime transport of current-use pesticides from California's Central Valley to the Sierra Nevada Mountain Range, USA. *Environ. Toxicol. Chem. Int. J.* 18, 2715–2722.

Lu, J. (2018). Federal judges finally just order the EPA to ban a dangerous pesticide. *Pop. Sci.* https://www.popsci.com/epa-federal-chlorpyrifos-ban/ (Accessed 2/1/2021).

McConnell, L.L., LeNoir, J.S., Datta, S., and Seiber, J.N. (1998). Wet deposition of current-use pesticides in the Sierra Nevada Mountain range, California, USA. *Environ. Toxicol. Chem. Int. J.* 17, 1908–1916.

Mueller, T.C., and Steckel, L.E. (2019). Dicamba volatility in humidomes as affected by temperature and herbicide treatment. *Weed Technol.* 33, 541–546.

National Pesticide Information Center. (2020). *Pesticides and Water Resources*. http://npic.orst.edu/envir/waterenv.html (Accessed 2/1/2021).

Odenkirchen, E.W., and Eisler, R. (1988). *Chlorpyrifos Hazards to Fish, Wildlife, and Invertebrates: A Synoptic Review*. Laurel, MD: Fish and Wildlife Service.

Oseland, E., Bish, M., Steckel, L., and Bradley, K. (2020). Identification of environmental factors that influence the likelihood of off-target movement of dicamba. *Pest Manag. Sci.* 76, 3282–3291.

Ross, L., Nicosia, S., McChesney, M., Hefner, K., Gonzalez, D., and Seiber, J. (1990). *Volatilization, Off-site Deposition, and Dissipation of DCPA in the Field. J. Environ. Qual.* 19(4), 715–722.

Seiber, J., and Woodrow, J. (1981). Sampling and analysis of airborne residues of paraquat in treated cotton field environments. *Arch. Environ. Contam. Toxicol.* 10, 133–149.

Seiber, J.N., Ferreira, G.A., Hermann, B., and Woodrow, J.E. (1980). Analysis of pesticidal residues in the air near agricultural treatment sites. In J. Harvey, Jr., and G. Zweig, (Eds.), *American Chemical Society Symposium Series No. 136 Pesticide Analytical Methodology*, Washington, D.C.: ACS Publications.

Seiber, J.N., McChesney, M.M., and Woodrow, J.E. (1989). Airborne residues resulting from use of methyl parathion, molinate and thiobencarb on rice in the Sacramento Valley, California. *Environ. Toxicol. Chem. Int. J.* 8, 577–588.

Siddaramappa, R., Tirol, A.C., Seiber, J., Heinrichs, E., and Watanabe, I. (1978). The degradation of carbofuran in paddy water and flooded soil of untreated and retreated rice fields. *J. Environ. Sci. Health Part B* 13, 369–380.

Smalling, K.L., Fellers, G.M., Kleeman, P.M., and Kuivila, K.M. (2013). Accumulation of pesticides in pacific chorus frogs (*Pseudacris regilla*) from California's Sierra Nevada Mountains, USA: Pesticides residues in amphibians. *Environ. Toxicol. Chem.* 32, 2026–2034.

Waite, D., Bailey, P., Sproull, J., Quiring, D., Chau, D., Bailey, J., and Cessna, A. (2005). Atmospheric concentrations and dry and wet deposits of some herbicides currently used on the Canadian Prairies. *Chemosphere* 58, 693–703.

Widder, P.D., and Bidwell, J.R. (2006). Cholinesterase activity and behavior in chlorpyrifos-exposed Rana sphenocephala tadpoles. *Environ. Toxicol. Chem. Int. J.* 25, 2446–2454.

Woodrow, J.E., Crosby, D.G., and Seiber, J.N. (1983). Vapor-phase photochemistry of pesticides. In F.A. Gunther, and J.D. Gunther, (Eds.), *Residue Reviews.* New York: Springer.

Woodrow, J.E., Gibson, K.A., and Seiber, J.N. (2018). Pesticides and related toxicants in the atmosphere. In P. de Voogt, (Ed.), *Reviews of Environmental Contamination and Toxicology* (Vol. 247, pp. 147–196). Cham: Springer International Publishing.

Woodrow, J.E., McChesney, M.M., and Seiber, J.N. (1990). Modeling the volatilization of pesticides and their distribution in the atmosphere. *Long Range Transp. Pestic.* Ed. Kurtz DA 61–81.

Woodrow, J.E., Seiber, J.N., and Baker, L.W. (1997). Correlation techniques for estimating pesticide volatilization flux and downwind concentrations. *Environ. Sci. Technol.* 21, 523–529.

Woodrow, J.E., Seiber, J.N., and Dary, C. (2001). Predicting pesticide emissions and downwind concentrations using correlations with estimated vapor pressures. *J. Agric. Food Chem.* 49, 3841–3846.

# 11

## Viruses, Pathogens, and Other Contaminants

### 11.1 Introduction

Aside from pesticides and other contaminants discussed in previous chapters, other contaminants in air are also high priority to the public and regulatory agencies, including airborne viruses and pathogens. Poor air quality is associated with at least 6 million premature deaths annually (Landrigan et al., 2018). Fine particulate matter (PM2.5) is the lung penetrating fraction that is associated with adverse health conditions, including respiratory disease and death. Regulations in developed countries have led to significant declines in PM2.5 and other pollutants; however, rapidly developing nations are experiencing high levels of particulate matter in ambient air and in some regions, levels are increasing due to the pressure to rapidly industrialize (Figure 11.1) (Landrigan et al., 2018).

From approximately January 2020 to June 2021, at least 3.5 million people died from the airborne severe acute respiratory syndrome coronavirus 2 (SARS-CoV-2), which caused the novel coronavirus disease of 2019 (COVID-19) worldwide (World Health Organization, 2021). There is a striking relationship between death from COVID-19 and air pollution. In the United States, counties with historically high levels of PM2.5 had higher mortality rates due to COVID-19: an increase in 1 $\mu g/m^3$ long-term exposure led to 11% increase in COVID-19 mortality rate (Wu et al., 2020).

### 11.2 COVID-19

In early January 2020, China reported 44 cases of pneumonia from unknown causes to the World Health Organization (2020). The initial cluster of cases was linked to a "wet market" in Wuhan, China. Shortly thereafter, a novel coronavirus was discovered to be the cause. It was ~80% identical to SARS-CoV and 96% identical to a bat coronavirus (Zhou et al., 2020). In the same study, they identified angiotensin-converting enzyme II (ACE2) as the

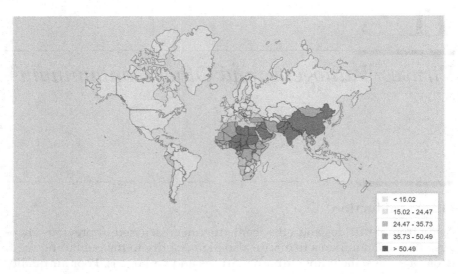

**FIGURE 11.1**
PM2.5 air pollution, mean annual exposure ($\mu g/m^3$), 2017 data. (Credit: Brauer, M. et al. (2017), Global Burden of Disease Study. The World Bank, https://data.worldbank.org/indicator/EN. ATM.PM25.MC.ZS accessed 03/16/2021, License: CC BY-4.0.)

receptor for entry to the cell (same as SARS-CoV). Symptoms in early cases included fever, upper and lower respiratory tract symptoms, or diarrhea, or a combination occurring 3–6 days after infection and the conditions were more severe in older patients (>65 years old) (Chan et al., 2020).

As of January 26, 2020, 2,014 confirmed cases of COVID-19 had been reported from China (1,985 cases), Thailand (5), Japan (3), the Republic of Korea (2), Vietnam (2), Singapore (4), Australia (4), Malaysia (3), Nepal (1), the United States (2), and France (3) (World Health Organization, 2020). According to the latest report from WHO (at the time of this writing), more than 2.7 million new COVID-19 cases were reported in the previous week and more than 60,000 new deaths mostly from the Americas and Europe (World Health Organization, 2021). The cases reported on March 9, 2021 were an increase in 2% compared to the week prior to that.

### 11.2.1 Treatments for COVID-19

As of March 16, 2021, there is no approved cure for COVID-19; however, promising leads are reported weekly, in *Science* and other outlets. The disease is currently treated with physical interventions, including turning patients to prone position and ventilation with ventilators, and pharmaceutical interventions with antivirals, immune system mimics, and immunomodulatory drugs. Despite the unprecedented development of new vaccines (described below), drugs are needed to treat patients while the world gets vaccinated.

During the worldwide vaccination campaign, many people are expected to become infected and may even die.

Antivirals stop or slow viral replication. Among the antivirals, remdesivir (Gilead Sciences, Inc.) is the only FDA approved treatment; however, its efficacy is controversial. A clinical trial conducted by WHO in October 2020 found no evidence that remdesivir reduces mortality, need for ventilation, or hospitalization duration and WHO recommends against remdesivir. Another clinical trial reported remdesivir is superior to the placebo (Beigel et al., 2020). Ivermectin, another antiviral, reportedly reduced the duration of illness (Ahmed et al., 2021); however, the U.S. National Institutes of Health (NIH) advises that there is not enough efficacy data to recommend for or against use of ivermectin (NIH, 2021).

New antivirals designed for COVID-19 are desperately needed. Repurposing previously approved drugs and artificial intelligence to screen billions of compounds are accelerating discovery of new drug leads (Service et al., 2021). Plitidepsin, approved to treat relapsed and refractory multiple myeloma, was 27.5 times more potent than remdesivir against SARS-CoV-2 in vitro, with limited toxicity in cell culture. Antiviral activity was mediated through inhibition of host protein (eEF1A, eukaryotic translation elongation factor 1A) (White et al., 2021). Targeting a host protein makes it unlikely the virus could evolve to evade drugs. Dai et al. reported in a cover article in *Science* (Dai et al., 2020) design and synthesis of two lead compounds (11a and 11b) that target M pro, a protease that plays a pivotal role in mediating viral replication and transcription in SARS-CoV-2. Both compounds showed good pharmacokinetic properties in vivo, and 11a also exhibited low toxicity, which suggests that these compounds are promising drug candidates.

Immune system mimics are another treatment. Convalescent plasma (from patients who already had COVID-19) gained a lot of attention, but most trials have ruled it out as an effective treatment (Katz, 2021). Perhaps the most promising news in treatment and prevention of COVID-19 (aside from vaccines) is the monoclonal antibody treatment bamlanivimab from Eli Lily. Injection of bamlanivimab gives immediate protection, reducing new COVID-19 infections by 80%. This drug is useful in preventing outbreaks in nursing homes (Eli Lilly and Company, 2021; Kolata, 2021).

Finally, there are treatments that reduce the immune system's response. Among these, the corticosteroid, mostly dexamethasone, is widely used and is thought to reduce the death rate (The RECOVERY Collaborative Group, 2020). Another emerging immunomodulatory strategy addresses the "cytokine storm" which is thought to be the cause for the most severe cases that lead to death. Some cytokine inhibitors have been associated with reduction in risk of death, but as with most treatments, more research is needed (Khan et al., 2021). A drug normally used for treatment of obsessive-compulsive disorder, fluvoxamine, was studied in a small clinical trial. The initial idea came from Dr. Angela Reiersen. While lying on the couch and recovering from COVID-19 herself, she remembered that the drug had proven effective against sepsis (a runaway immune response) caused by bacterial infections. Reiersen and coworkers hypothesized the drug could be used as an early treatment for those testing positive for COVID-19 to prevent severe systems caused by the immune system (Alfonsi, 2021; Kabani, 2021).

### 11.2.2 Vaccines for COVID-19

As of this writing, there are at least 89 vaccine candidates under development and 2 approved for full use in the United States, the European Union, and other places. The leading vaccines as of March 16, 2021 are reported in Table 11.1 (source: NYTimes Coronavirus vaccine tracker (Zimmer et al., 2020)).

### 11.2.3 New Variants of COVID-19

Despite the unprecedented speed at which vaccines have been developed for COVID-19, they are being rolled out painfully slow and there are widespread shortages. The longer it takes to slow the spread and reach herd immunity, the more risk there is of new variants of the virus to develop and spread. New variants are complicating the current situation, and vaccine developers are rushing to evaluate the efficacy of vaccines against new variants. According to the U.S. Centers for Disease Control (CDC), as of January 15, 2021, it is not known how widely these variants have spread, how the disease differs from older strains, and efficacy of the current therapies and vaccines:

- B.1.1.7: First detected in the United Kingdom (UK) on September 20, 2020, it had key mutations including deletion of H69/V70; deletion of Y144; and mutation of N501Y, A570D, and P681H. It has increased transmissibility and possible increased risk of hospitalization, severity, and mortality. Currently approved vaccines may have lower efficacy with this strain: Moderna, Pfizer-BioNTech had reduced neutralizing activity (i.e., lower levels of functional systemic antibodies to prevent infectivity of the virus), but the impact on protection against the disease is not currently known. The Oxford-AstraZeneca vaccine had limited efficacy against mild–moderate COVID-19. A small sample size indicated substantially lower neutralization against this strain relative to the original strain. It has been reported in 58 different countries. As of March 11, 2021, 4,690 cases have been reported in 50 jurisdictions in the United States (CDC, 2020).

- B.1.351: First detected in South Africa in early August of 2020 had key mutations including deletion of L242/A243/L244, and mutation of K417N, E484K, and N501Y. It has increased transmissibility, but no known impacts on severity. It potentially has higher risk for reinfection. It does not appear to impact the efficacy of Moderna,

**TABLE 11.1**

Several Vaccines Approved for COVID-19

| Vaccine | Type | Administered | Efficacy (%) | Storage | Authorized Use |
|---|---|---|---|---|---|
| Pfizer, BioNTech | mRNA | 2 doses, 3 weeks apart | 95 | −25°C to −15°C | Emergency use approved in the United States (December 11, 2020) and fully approved in other countries (Bahrain, Brazil, New Zealand, Saudi Arabia, Switzerland) |
| Moderna | mRNA | 2 doses, 4 weeks apart | 94.5 | 30 days with refrigeration, 6 months at −20°C | Emergency use approved in the United States (December 18, 2020) and fully approved in Switzerland |
| Johnson & Johnson | Adenovirus-based (Ad26) | 1 dose | Up to 72 | Up to 2 years frozen (−20°C), and up to 3 months refrigerated (2°C–8°C). | Emergency use approved in the United States (February 27, 2021) and other countries |
| Gamaleya Research Institute, Sputnik V | Adenovirus-based (Ad26, Ad5) | 2 doses, 3 weeks apart | 91.6 | Freezer storage | Approved for early use in Russia (December 14, 2020). Emergency use in Belarus, other countries |
| Oxford-AstraZeneca, AZD1222 | Adenovirus-based (ChAdOx1) | 2 doses, up to 12 weeks apart | Up to 82.4 | Stable in refrigerator for at least 6 months | Emergency use in Britain, India, other countries (Dec. 30, 2020). Denmark, Iceland and Norway (March 11), Germany, France, and Italy (March 15) suspended the use of the vaccine because of concerns about a possible increased risk of blood clots |
| CanSino Biologics, Convidecia | Adenovirus-based (Ad5) | Single dose | 65.28 | Refrigerated | Approved in China, emergency use in Mexico and Pakistan |
| Sinopharm, BBIBP-CorV | An inactivated coronavirus vaccine | 2 doses, 3 weeks apart | 79.34 | | Approved in China, the United Arab Emirates, Bahrain. Emergency use in Egypt, Iraq, Nepal, Peru, and other countries |
| Sinovac, CoronaVac | Inactivated vaccine | 2 doses, 2 weeks apart | Up to 83.5 | Refrigerated | Approved in China, Emergency use in Azerbaijan, Brazil, Mexico, Ukraine, and others |
| Bharat Biotech, Covaxin | Inactivated form of the coronavirus | 2 doses, 4 weeks apart | 80.6 | At least a week at room temperature | Emergency use in India, Iran, and Zimbabwe |

*Source:*   Data from NYTimes Coronavirus vaccine tracker (Zimmer et al., 2020). Updated March 15, 2021.

Pfizer-BioNTech, and Oxford-AstraZeneca vaccines. It has been reported in 111 different countries. As of March 11, 2021, 143 cases have been reported in 25 jurisdictions in the United States (CDC, 2020).

- B.1.1.28.1, also known as P.1: First detected in Brazil and Japan in December of 2020 had key mutations of K417N, E484K, and N501Y. It has increased transmissibility, and impacts on severity are under investigation. Reinfections have been reported and impacts on vaccines are not known. It has been reported in 32 different countries. As of March 11, 2021, 25 cases have been reported in 10 jurisdictions in the United States (CDC, 2020).

### 11.2.4 Transmission of COVID-19

Airborne viruses and fine particulate matter have remarkable parallels in how they are transmitted and accumulate in the respiratory tract. The coronavirus is similar in size to fine particulate matter. Infected people, many asymptomatic, produce respiratory droplets and aerosols carrying the virus that are transmitted in close quarters or in unventilated spaces, enabling it to be inhaled into deeper air passages in the lungs. Just as fine particles from cigarette smoke or pollution can be retained in the lungs, leading over time to a range of adverse health effects including chronic lung diseases such as asthma and bronchitis, as well as cancer; coronavirus attacks the lungs (as well as other organs) once it reaches a critical level of exposure, triggering an often-fatal inability to breathe. In both cases, airborne transmission and persistence likely play a critical role.

A potential breakthrough that supports the case for transmission of COVID-19 via respiratory droplets, or aerosols, occurred with the preliminary publication of research conducted at the University of Florida in August 2020 (Lednicky et al., 2020), where a team "succeeded in isolating live virus from aerosols collected at a distance of seven to 16 feet from patients hospitalized with COVID-19—farther than the six feet recommended in social distancing guidelines." In the new study, researchers devised a sampler that uses pure water vapor to enlarge the aerosols enough that they can be collected easily from the air. Rather than leave these aerosols sitting, the equipment immediately transfers them into a liquid that is rich with salts, sugar, and protein, which preserves the pathogen. It is a tantalizing clue toward understanding the role that aerosol-based transmission plays in the spread of the contagion.

Other studies provide strong evidence that airborne transmission of COVID-19 follows the same basic principles of transmission of other small size airborne contaminants. An analysis of a call center in Seoul, South Korea, showed how a coronavirus outbreak traveled among workers in one half of the floor of a building in a tight, shared airspace with no outside ventilation,

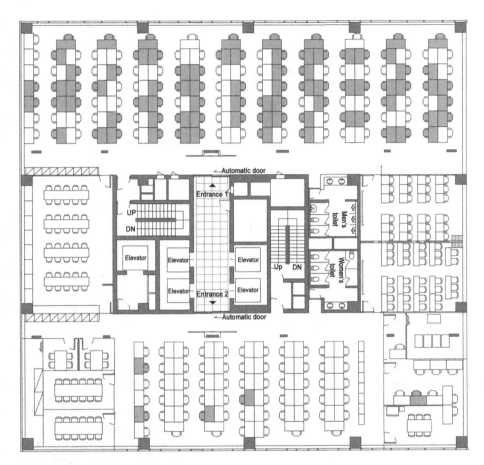

**FIGURE 11.2**
Shared indoor space. (Reprinted from Park, S.Y., Kim, Y.-M., Yi, S., Lee, S., Na, B.-J., Kim, C.B., Kim, J., Kim, H.S., Kim, Y.B., Park, Y., et al. (2020, August). Coronavirus disease outbreak in call center, South Korea. *Emerg. Infect. Dis. J.* – CDC, 26(8). https://wwwnc.cdc.gov/eid/article/26/8/20-1274_article Courtesy of the Centers for Disease Control and Prevention.)

while workers in the other half of the floor remained largely virus-free (Figure 11.2) (Park et al., 2020).

In another study, researchers at the Guangzhou Center for Disease Control and Prevention investigated a COVID-19 outbreak traced to a restaurant in Guangzhou, China (Lu et al., 2020). One asymptomatic person (A1) appeared to have infected nine other people as a result of being in the shared air path of an air conditioner (Figure 11.3). Other patrons outside of the path of the airflow, and restaurant staff, did not get sick. "We conclude that in this outbreak, droplet transmission was prompted by air-conditioned ventilation," wrote the researchers.

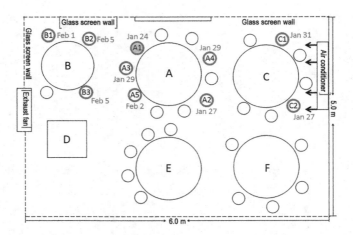

**FIGURE 11.3**
A floor plan of restaurant tables and air-conditioning flow at the site of a coronavirus outbreak in Guangzhou, China. The yellow circle indicates an infected patient; red circles show future patients (Reprinted from Lu, J., Gu, J., Li, K., Xu, C., Su, W., Lai, Z., Zhou, D., Yu, C., Xu, B., and Yang, Z. (2020, July). COVID-19 outbreak associated with air conditioning in restaurant, Guangzhou, China. *Emerg. Infect. Dis. J.* - *CDC*. 26(7). https://dx.doi.org/10.3201/eid2607.200764. Courtesy of the Centers for Disease Control and Prevention.)

Additional research has attempted to model the spread of COVID-19 via air in environments (Azimi et al., 2020). "A research team based at Harvard and the Illinois Institute of Technology tried to understand how the virus moved from person to person and room to room on the cruise ship Diamond Princess, in which 700 of 3,711 passengers tested positive with the virus in a month," (Carey and Glanz, 2020). Their computer model indicated that the virus spread most likely via microscopic droplets light enough to linger in the air and circulate via the ship's ventilation system. The study reinforced use of efficient masks, social distancing, and frequent replenishment of ventilated air with fresh air as important tools to prevent virus spread.

## 11.3 Other Airborne Diseases

Coronaviruses are one of many classes of diseases that can be transmitted through air. For example, coccidioidomycosis is a fungal infection caused by inhaling the spores of either *Coccidioides immitis* or *Coccidioides posadasii* fungi. These spores are endemic to the soils of the American Southwest and can get into the air when the soil is disturbed—for example, via wind, construction, gardening, or farming (American Thoracic Society, 2011). Like coronaviruses, the severity of the infection is largely dependent on the degree of exposure to the spores, along with the health of the victim's immune system. A severe infection can lead to "Valley Fever," a dangerous and sometimes fatal disease (as one of the authors experienced firsthand while a graduate student at Arizona State University and Utah State University). Another disease that can be transmitted through air is Legionnaires' disease—a severe form of pneumonia. Caused by the legionella bacterium, Legionnaires' disease comes from inhaling the bacteria from water or soil (Mayo Clinic, 2019). As suspected with COVID-19, the degree of exposure is correlated with the severity of the illness. In the summer months, malfunctioning air conditioning equipment can be a breeding ground for the bacteria and lead to prolonged exposure in hotels in the Southwest—much as air conditioning appears to be a mechanism for contaminating an airspace with coronaviruses.

Coronavirus is a zoonotic disease, inviting further comparison to pathogenic fungi commonly associated with food poisoning. Pathogen-based food poisoning is most often associated with bacteria such as *E. coli*, *Salmonella*, *Campylobacter*, and others in contaminated lettuce, sprouts, and other leafy greens, as well as in almonds and other tree nut crops. It is primarily spread in improperly washed, processed, or stored foods. However, there are reports of airborne transport of pathogens like *E. coli* from animal herds and feed lots to downwind vegetable fields and orchards. This may be exacerbated by the presence of animal feces in and around fields adjacent to vegetable growing areas such as in the San Joaquin and Salinas Valleys (Harris, Mandrell, UC, Davis, personal communication). Manure also creates particulate matter containing pathogens like fungi that get distributed in air or get into the food supply, leading to sickness and fatalities. Released or escaped domestic pigs can also contaminate vegetable fields through feces, leading to *E. coli* outbreaks. The European boar, introduced possibly by Spanish explorers and settlers, causes $800 million in estimated damage in the United States and is present in ~40 states. And, it carries more than 30 viral and bacterial diseases.

Inhalation of anthrax spores (*Bacillus anthracis*), the most dangerous type of exposure, causes flu-like symptoms (fevers, chills, stomach pains, body aches, etc.) and is accompanied by a drenching sweat (U.S. CDC). Without treatment mortality is close to 90%, but drops to 55% when aggressively treated. People can be exposed when working with infected animals or by products obtained from infected animals.

Anthrax is a danger due to its facile development into a biological weapon: it is naturally occurring, easy to culture, easily dispersed, and persists in the environment. Anthrax was used in a bioterrorism attack in 2001: letters containing anthrax were mailed to several U.S. officials. Of the 22 exposures, five people died (Ember, 2006). In a later incident, Dugway Proving Ground (U.S. military facility) accidentally sent live anthrax spores through the mail to U.S. labs and a lab in South Korea. The mistake was blamed on poor oversight and the difficulty in using irradiation to deactivate spores (Reardon, 2015).

## 11.4 New Insights into Viral Transmission

There are of course many other diseases transmitted by airborne contact, ranging from colds to influenzas to hantavirus, other airborne microorganisms, plagues, and tuberculosis. There is also a broad array of plant viruses, some of which are spread in air. Common to many of these is a lack of knowledge about their transport, fate, and persistence in air. The coronavirus pandemic is serving to accelerate research into airborne transmission of diseases, opening up new avenues for discovery. A study conducted by the University of California, Davis, and the Icahn School of Medicine at Mt. Sinai suggests that viruses may spread through the air on dust, fibers, and other microscopic particles (Asadi et al., 2020)—a particular problem given the wildfires ravaging the western United States at the time of this writing. Another study conducted by researchers from Yangzhou University in China suggested that flushing a toilet can release clouds of virus-laden aerosols (Wang et al., 2020). And research in Australia appears to suggest that low humidity is correlated with greater spread of COVID-19 (Ward et al., 2020).

Pesticide research and assessment principles/techniques could apply to studying/modeling the transmission of COVID-19, i.e., measurement, drift, and intervention. The U.S. EPA and California Department of Pesticide Regulation are involved in registration of control agents for viruses. These chemicals are registered pesticides (https://www.cdpr.ca.gov/docs/-covid-19/). To date there is a lack of cross talk between public health personnel and agricultural and environmental scientists and engineers. It may take a pandemic or other catastrophe to promote interdisciplinary cooperation!

## References

Ahmed, S., Karim, M.M., Ross, A.G., Hossain, M.S., Clemens, J.D., Sumiya, M.K., et al. (2021). A five-day course of ivermectin for the treatment of COVID-19 may reduce the duration of illness. *Int. J. Infect. Dis.* 103, 214–216.

Alfonsi, S. (2021). Finding a possible early treatment for COVID-19 in a 40-year-old antidepressant. CBS News https://www.cbsnews.com/news/fluvoxamine-antidepressant-drug-covid-treatment-60-minutes-2021-03-07/ (Accessed 03/16/2021).

American Thoracic Society (2011). Coccidioidomycosis. *Am. J. Respir. Crit. Care Med.* 184, 5–6.

Asadi, S., Gaaloul ben Hnia, N., Barre, R.S., Wexler, A.S., Ristenpart, W.D., and Bouvier, N.M. (2020). Influenza A virus is transmissible via aerosolized fomites. *Nat. Commun.* 11, 4062.

Azimi, P., Keshavarz, Z., Laurent, J.G.C., Stephens, B.R., and Allen, J.G. (2020). Mechanistic transmission modeling of COVID-19 on the diamond princess cruise ship demonstrates the importance of aerosol transmission. *MedRxiv* 2020.07.13.20153049.

Beigel, J.H., Tomashek, K.M., Dodd, L.E., Mehta, A.K., Zingman, B.S., Kalil, A.C., et al. (2020). Remdesivir for the treatment of Covid-19 — final report. *N. Engl. J. Med.* 383, 1813–1826.

Brauer, M. et al. (2017), Global Burden of Disease Study. The World Bank, https://data. worldbank.org/indicator/EN.ATM.PM25.MC.ZS (Accessed 03/21/2021).

Carey, B., and Glanz, J. (2020). Aboard the Diamond Princess, a case study in aerosol transmission. *N. Y. Times.*

CDC. (2020). COVID-19 and your health. https://www.cdc.gov/coronavirus/2019-ncov/transmission/variant.html (Accessed 01/16/2021).

Chan, J.F.-W., Yuan, S., Kok, K.-H., To, K.K.-W., Chu, H., Yang, J., et al. (2020). A familial cluster of pneumonia associated with the 2019 novel coronavirus indicating person-to-person transmission: a study of a family cluster. *The Lancet* 395, 514–523.

Dai, W., Zhang, B., Jiang, X.-M., Su, H., Li, J., Zhao, Y., et al. (2020). Structure-based design of antiviral drug candidates targeting the SARS-CoV-2 main protease. *Science* 368, 1331–1335.

Eli Lilly and Company. (2021). A phase 3 randomized, double-blind, placebo-controlled trial to evaluate the efficacy and safety of LY3819253 alone and in combination with LY3832479 in preventing SARS-CoV-2 infection and COVID-19 in skilled nursing and assisted living facility residents and staff; a NIAID and Lilly collaborative study (clinicaltrials.gov).

Ember, L. (2006). Anthrax sleuthing. https://cen.acs.org/articles/84/i49/Anthrax-Sleuthing.html.

Kabani, M. (2021). COVID-19 research points to repurposed drugs. CBS News. https://www.cbsnews.com/news/covid-19-repurposed-drugs-60-minutes-2021-03-07/ (Accessed 03/16/2021).

Katz, L.M. (2021). (A Little) clarity on convalescent plasma for Covid-19. *N. Engl. J. Med.*

Khan, F., Stewart, I., Fabbri, L., Moss, S., Robinson, K.A., Smyth, A., and Jenkins, G. (2021). A systematic review and meta-analysis of Anakinra, Sarilumab, Siltuximab and Tocilizumab for Covid-19. *MedRxiv* 2020.04.23.20076612.

Kolata, G. (2021). Drug prevents coronavirus infection in nursing homes, maker claims. *N. Y. Times.*

Landrigan, P.J., Fuller, R., Acosta, N.J.R., Adeyi, O., Arnold, R., Basu, N., et al. (2018). The Lancet commission on pollution and health. *The Lancet* 391, 462–512.

Lednicky, J.A., Lauzardo, M., Fan, Z.H., Jutla, A., Tilly, T.B., Gangwar, M., et al. (2020). Viable SARS-CoV-2 in the air of a hospital room with COVID-19 patients. *Int. J. Infect. Dis.* 100, 476–482.

Lu, J., Gu, J., Li, K., Xu, C., Su, W., Lai, Z., Zhou, D., Yu, C., Xu, B., and Yang, Z. (2020, July). COVID-19 outbreak associated with air conditioning in restaurant, Guangzhou, China. *Emerg. Infect. Dis. J. - CDC.* 26(7).

Mayo Clinic. (2019). Legionnaires disease. https://www.mayoclinic.org/diseases-conditions/legionnaires-disease/symptoms-causes/syc-20351747#:~:text=Legionnaires'%20disease%20is%20a%20severe,bacteria%20from%20water%20or%20soil (Accessed 03/06/2021).

NIH. (2021). NIH COVID-19 treatment guidelines: Ivermectin. https://www.covid19treatmentguidelines.nih.gov/antiviral-therapy/ivermectin/ (Accessed 03/16/2021).

Park, S.Y., Kim, Y.-M., Yi, S., Lee, S., Na, B.-J., Kim, C.B., Kim, J., Kim, H.S., Kim, Y.B., Park, Y., et al. (2020, August). Coronavirus disease outbreak in call center, South Korea. *Emerg. Infect. Dis. J. – CDC*, 26(8), 1666–1670.

Reardon, S. (2015). US military accidentally ships live anthrax to labs. *Nature News.*

Service, R.F. (2021). Researchers race to develop antiviral weapons to fight the pandemic coronavirus. https://www.sciencemag.org/news/2021/03/researchers-race-develop-antiviral-weapons-fight-pandemic-coronavirus.

The RECOVERY Collaborative Group. (2020). Dexamethasone in hospitalized patients with Covid-19 — preliminary report. *N. Engl. J. Med.*, 384, 693–704.

US CDC Guide to Understanding Anthrax. 9. https://www.cdc.gov/anthrax/pdf/evergreen-pdfs/anthrax-evergreen-content-english.pdf (Accessed 03/16/2021).

Wang, J.-X., Li, Y.-Y., Liu, X.-D., and Cao, X. (2020). Virus transmission from urinals. *Phys. Fluids* 32, 081703.

Ward, M.P., Xiao, S., and Zhang, Z. (2020). Humidity is a consistent climatic factor contributing to SARS-CoV-2 transmission. *Transbound. Emerg. Dis.* 67, 3069–3074.

White, K.M., Rosales, R., Yildiz, S., Kehrer, T., Miorin, L., Moreno, E., et al. (2021). Plitidepsin has potent preclinical efficacy against SARS-CoV-2 by targeting the host protein eEF1A. *Science* 371, 926–931.

World Health Organization. (2020, January 26). WHO weekly update. https://www.who.int/publications/m/item/weekly-operational-update-on-covid-19---26-january-2021 (Accessed 03/16/2021).

World Health Organization. (2021, March 9). WHO weekly update. https://www.who.int/publications/m/item/weekly-epidemiological-update---10-march-2021 (Accessed 03/16/2021).

Wu, X., Nethery, R.C., Sabath, M.B., Braun, D., and Dominici, F. (2020). Air pollution and COVID-19 mortality in the United States: Strengths and limitations of an ecological regression analysis. *Sci. Adv.* 6, eabd4049.

Zhou, P., Yang, X.-L., Wang, X.-G., Hu, B., Zhang, L., Zhang, W., et al. (2020). A pneumonia outbreak associated with a new coronavirus of probable bat origin. *Nature* 579, 270–273.

Zimmer, C., Corum, J., and Wee, S.-L. (2020). Coronavirus Vaccine Tracker. *N. Y. Times.*

# 12

## Biopesticides and the Toolbox Approach to Pest Management

### 12.1 Introduction

Pesticides may be defined as chemicals (or mixtures of chemicals) used to restrict or repel pests such as insects, weeds, fungi, nematodes, and other organisms that adversely affect food production, ecosystem function, or human health. Pesticides may also be toxic contaminants in our food supply and environment (air, water, soil), and responsible for illness or injury to people and wildlife. Biopesticides are attracting interest as alternatives to conventional pesticides, but without many of the nontarget effects, by offering improved safety in pest control practices. In this chapter, we summarize and discuss the current status and future promise of biopesticides, including in China and other developing agricultural economies, and how biopesticides use may increase the safety and sustainability of the food supply.

### 12.2 Current Trends

Pesticide use in California and the United States peaked in the early 1980s. Since then, there has been a trend toward less quantities of pesticide use (California Department of Pesticide Regulation, 2017). This trend reflects a combination of several factors: the banning or phaseout of high volume use synthetics, such as toxaphene, chlordane, and methyl bromide; development of more efficient application technology which delivers more chemical to the target and allows less chemical loss by volatilization. wind erosion, and by surface runoff; and the introduction of transgenic modifications in some crops such as cotton, corn, and soybeans so that resistance or tolerance to pests is carried by the crop without need for, or minimal use of, external chemical application for pest control. An example is the technology underlying the use of transgenic *Bacillus thuringiensis* (Bt) toxin to control insects.

Farmers are also using more integrated pest management tools, such as inter-cropping, cover crops, biocontrol, and crop rotation, along with reduced risk chemicals such as synthetic pyrethroids, avermectins, and spinosad that are generally effective at lower application rates than conventional pesticides (Science, 2013). These tools all work to reduce the amount of chemical applied to crops to obtain economically acceptable levels of pest control. They will also improve the safety of food production by eliminating or minimizing the use of costly chemicals that leave toxic residues in foods, or lead to illness in farm worker populations, and do so at lower costs in many cases.

## 12.3 Biopesticides

"Biopesticides" would include all of the things in the definition of pesticides but with several modifications, in particular that they are naturally occurring or derived from natural products by straightforward chemical modification. The EPA definition is that biochemical biopesticides are natural compounds or mixtures that manage pests without a toxic mode of action, unlike cholinesterase inhibition which is a characteristic of most organophosphate and carbamate insecticides (Leahy et al., 2014).

Third-generation pest control agents, reduced risk pesticides, and biobased pesticides are other terms sometimes used in place of "biopesticides." The common elements of biopesticides can include some or all of the following: naturally occurring, reduced toxicity to nontarget organisms, low persistence in the environment or in ecological food chains, compatibility with organic farming methods, and low mammalian toxicity so that they are safe to handle, and not restricted in use by regulatory agencies (Cantrell et al., 2012; Ritter, 2012). Few products will fit all of these criteria, but the intent is clearly to stimulate "green" environmentally benign technologies for sustainable pest management and control.

Although the market is growing for biopesticides, none of the top-use pesticides in recent years in California or the United States clearly meet the biopesticide definition (California Department of Pesticide Regulation, 2017). One could argue that sulfur is close, since it is naturally occurring and of low nontarget toxicity, and useable in organic farming, but sulfur has been used for pest control for centuries and is not among the new generation of pest control agents (Griffith et al., 2015). Furthermore, sulfur must be refined from its natural sources in the environment before it can be marketed and sold for use as a pesticide. Various mineral oils used for weed control, some plant essential oils (e.g., orange oil for termite control), and corn gluten for weed control might be considered within the realm of biopesticides as well.

Basil (*Ocimum basilicum*) is a popular culinary herb throughout the world. Essential oil derived from basil (basil oil) has shown to be a potent biopesticide and food preservative (Li and Chang, 2016). Basil oil contains more than 200 chemicals such as linalool, estragole, and eugenol. The U.S. Food and Drug Administration (FDA) and European Commission have accepted and classified the components from basil (*O. basilicum*) oil as generally recognized as safe (GRAS). Carvacrol, linalool, and pulegone showed excellent potency against thrips, melon fly, and turnip aphids (Sampson et al., 2005). Basil oil and its major components trans-anethole, estragole, and linalool are very potent against adult fruit flies *Ceratitis capitata, Bactrocera dorsalis*, and *Bactrocera cucurbitae* (Chang et al., 2009). Electrophysiological studies indicated that linalool inhibits both mammalian γ-aminobutyric acid type A receptor (GABAAR) and nicotinic acetylcholine receptor (nAChR) (Li et al., 2020), which may explain its insecticidal activity. Linalool is a concentration-dependent, noncompetitive inhibitor on the GABAAR. The half maximal inhibitory concentration ($IC_{50}$) of linalool on the GABAAR and nAChR was approximately 3.2 and 3.0 mM, respectively (Li et al., 2020).

Spinosad, a combination of two natural insecticides, well represents the commercial possibilities for biopesticides, recently gaining a large market share for protection of apples, pears, strawberries, and other high-value crops. The residues left by spinosad are of low toxicity, and produce is considered safe for consumers, including infants and children, when the product is applied in the manner specified on the label. In the United States, spinosad has been approved for organic production. Spinosad is composed of two spinosyn active forms, A and D, differing only by placement of a methyl group (see Chapter 2, Figure 2.10). They are produced by soil-borne fungi (*Saccharopolyspora spinosa*) which can also be used in fermentation culture to produce the technical product. The use of fermentation to produce chemical control agents is another potential advantage for biopesticides since it can reduce or eliminate the need for extensive chemical processing facilities associated with pesticides based on petrochemicals.

In addition to Dow's spinosad, Merck and other firms have developed avermectins (Figure 12.1), macrocyclic lactones produced by fermentation of naturally occurring soil bacteria (*Streptomyces avermitilis*) which are useful for crop protection as well as treatment of livestock and pets for parasite and disease control.

A number of specialty pesticide companies, such as Trécé, Certis USA, and Marrone Bio Innovations (Table 12.1), have marketed new biopesticides, and now larger companies such as Corteva Agriscience™, Bayer, and BASF are developing and/or marketing biopesticides along with conventional (second generation) pest control chemicals. EPA has helped move the biopesticide technology forward by offering a "fast track" for registration of reduced toxicity pesticides. The newer bioproducts include fungicides, insect repellants and attractants (semiochemicals), insecticides, nematicides, herbicides, and products from genetic manipulation. In China, India, and several other developing economies, technologies for pest control are being developed based upon farmers' experience in those countries. In the Philippines, for

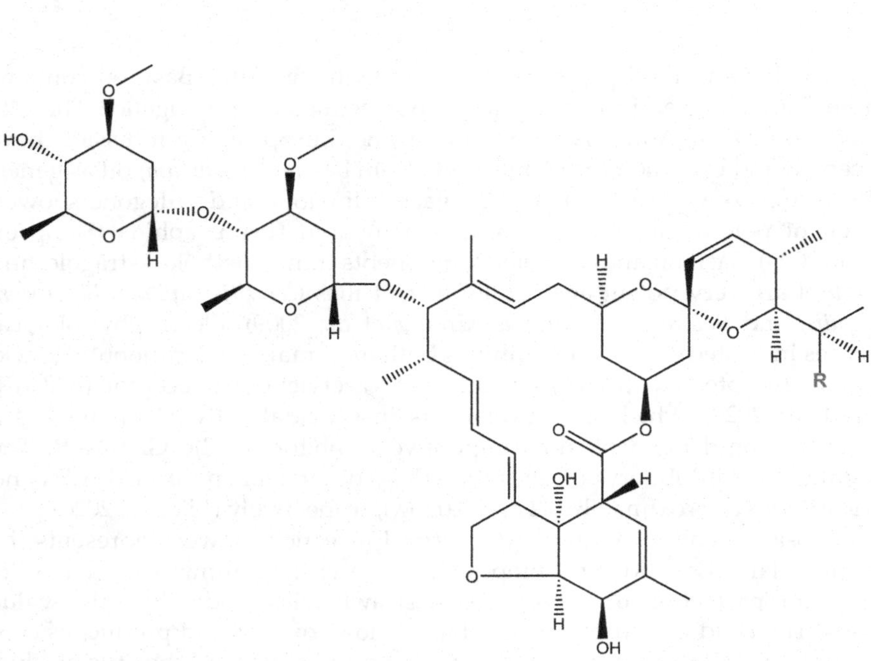

**FIGURE 12.1**
Avermectins are natural pesticides.

**TABLE 12.1**

Examples of Biopesticides Marketed by Specialty and Major Crop Protection Companies (Seiber et al., 2018)

| Biopesticide | Company | Use | Type |
|---|---|---|---|
| Spinosad | Dow | Insecticides | Spinosad and Spinosyns A and D |
| Avermectins | Merck and others | Anthelmintics, insecticides | Macrocyclic lactones |
| Requiem, Sonata, Ballad | Bayer Crop Science | Insecticide, fungicides | Microbial strains and mixtures |
| Serenade | Bayer Crop Science | Fungicide | Extract of giant knotweed |
| Cidetrak | Trécé | Insect control via mating disruption; gustatory stimulation coupled with an insecticide | Pheromones, kairomones |
| Venerate, Grandevo, Majestene | Marrone Bioinnovations | Insecticides, acaricides, nematicides | Microbial strains and mixtures |
| PFR-97, CYD-X, Gemstar, etc. | Certis USA | Insecticides, miticides | Insecticidal microbes and viruses |

*Source:* Seiber, J.N., Coats, J., Duke, S.O., and Gross, A.D. (2018). Pest management with biopesticides. *Front. Agri. Sci. Eng.* 5(3): 295–300. © The Author(s) 2018. Published by Higher Education Press. (http://creativecommons.org/licenses/by/4.0).

example, rice can be grown along with tilapia in the paddies. The tilapia fish consume insect pests of rice, so that less chemical control may be needed, and then become a nutritious coproduct harvestable along with the rice crop.

Bioherbicides are a future target for development, although research in herbicide discovery has been somewhat slow since the last new mode of action herbicides was introduced about 30 years ago (Duke, 2012). The evolution of weeds resistant to the leading herbicide glyphosate may accelerate developments in this area (Heap and Duke, 2018). Nonsynthetic chemical management of weeds in organic culture is a serious problem, limiting wider use of organic farming methods. The few bio- or green products for weed control use high application rates or multiple applications and even then are somewhat unpredictable in efficacy (Dayan and Duke, 2010). Mechanical control methods include hoeing, mowing, burning, or solarization under plastic sheeting tarps. Biocontrol of weeds with grazing sheep and goats has even been used. Many of these mechanical and biocontrol methods are laborious, costly, and often unreliable, or accompanied by undesirable environmental side effects.

Development of triketone herbicides is a case of a natural pesticide leading to synthesis of highly effective synthetic pesticides. Observation of herbicidal activity (allelopathy) of an ornamental plant (bottlebrush) led to the isolation of the naturally occurring bioactive principle, leptospermone (Figure 12.2), and various synthetic analogs that are now commercial herbicides such as mesotrione (Figure 12.3). Mesotrione was brought to market by Syngenta in 2001. It is a synthetic analog of leptospermone which mimics the herbicidal effects of this natural product (Dayan et al., 2007). It is a member of a class of inhibitors that work by inhibiting 4-hydroxyphenylpyruvate dioxygenase (HPPD). HPPD is required by plants for carotenoid and plastoquinone biosynthesis; carotenoids protect chlorophyll from sunlight-induced degradation and plastoquinone is required for photosynthesis. When the HPPD inhibitor is present in plants, carotenoids are prevented from being made and photosynthesis is inhibited, causing chlorophyll to degrade, followed by plant death. Sales by Syngenta were more than $400 million per year in 2011, but expiration of patents beginning in 2012 has opened the market to other synthetic triketone herbicides (Wikipedia, 2021).

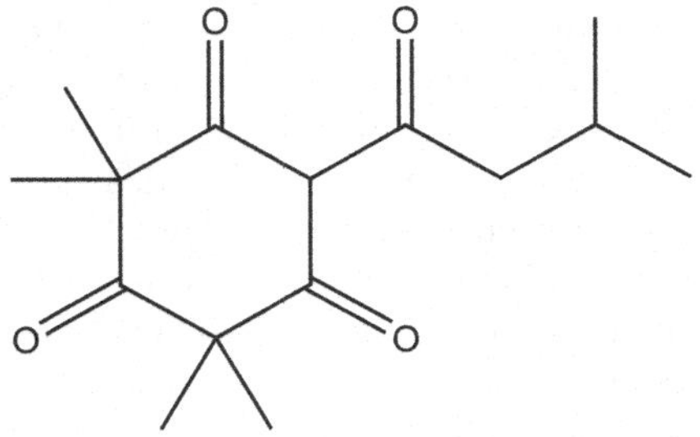

**FIGURE 12.2**
Leptospermone is a natural pesticide.

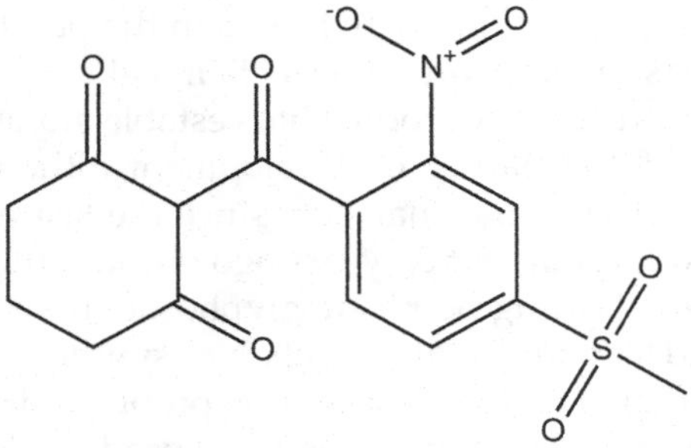

**FIGURE 12.3**
Mesotrione is a natural pesticide derivative.

Another future major market to be addressed is for bionematicides for soil application and use in stored products, given the mandated (Montreal Protocol (United Nations Environment Programme, 2018)) phaseout of methyl bromide, and off-target movement and exposure issues with present fumigants like methyl isothiocyanate (MITC), formed from synthetic metam products, and chloropicrin (see Chapter 8). Avermectins show some promise for nematode control, particularly in combination with other control measures, and farmers are adopting cultural methods (crop rotation, intercropping with Brassica species, solarization, and use of nematode resistant crops, etc.) to address nematodes in the more susceptible crops such as strawberries and carrots.

There is renewed interest in discovering botanical and related pesticides with novel structures and activity that can be used directly, or to inspire synthetic modification to form the pest control agents of the future. As Isman and others (Dayan et al., 2012; Isman and Grieneisen, 2014) have pointed out, plants produce a bewildering array of secondary metabolites thought to play an ecological role in defending plants from attack by herbivores and pathogens, as well as in chemically inhibiting competing plant species. These range from the familiar biopesticides of long-standing interest, such as pyrethrins, rotenones, and alkaloids, to complex mixtures of terpenes, carbohydrates, and proteins. Natural pesticide discovery is particularly pursued in Asia and Latin America as a way to overcome costly regulatory requirements in industrialized nations. The products are often complex unrefined mixtures of active ingredients and other components of presently unknown utility to the source plant or microbes.

Pesticide discovery has experienced renewed interest ranging from exploration of plants and microbes that produce chemicals and mixtures of potential utility to mainstream agriculture, to niche products that can be exploited by synthetic and biotechnological modifications in the future. There are plenty of areas yet to be explored, by chemical synthesis, biotechnology, and breeding, as well as by chemical ecologists and engineers. The success of the current group of biopesticides will likely be expanded as more scientists and practitioners are attracted to the critically important field of pest control in sustainable food production.

## 12.4 Other Alternatives to Synthetic Pesticides

The excitement over biopesticides, semiochemical communication cues, and other alternative controls, which was evident in the 1970s when the third-generation pest control movement was launched (Williams, 1967), is now regaining momentum. Over half of the new registrations for pesticides and pest control agents at EPA are for products associated with the features of biopesticides, and the market share is growing for these products (Cantrell et al., 2012).

High-throughput screening methods, nano-based encapsulation methods, and further development of semiochemicals for both monitoring and population control of pests offer promise for further developments. Semiochemicals are already far along in crop protection applications. Pheromones or synthetic analogs are widely used to survey for pest populations so that insecticide applications can be timed and positioned to be most effective. Mass trapping or confusion approaches have also been used with some success, using pheromones or synthetic or naturally occurring alternatives that disrupt pest insect populations. An example is the pheromone and a naturally occurring alternative with pheromone-like attractant activity found in pear leaves that can aid in control of codling moth in apple, pear, walnut, almond, and other crops susceptible to economic damage by codling moth (Light and Beck, 2010). Controlling this damaging pest, and other boring insects that affect cotton seed and peanuts, is a critical element in controlling invasion of Aspergillus fungi, which can infect pome fruit, nuts, or seeds and produce aflatoxins—a group of carcinogenic fungal metabolites. A combination of aflatoxin in foods and a hepatitis B-susceptible human population such as exists in many parts of Africa and Asia can lead to high levels of liver cancer and elevated mortality—a major public health and food safety concern. This is an example of how the use of effective pest control carries with it the added benefit of reducing the carcinogen load due to toxic natural products such as aflatoxins in the food supply—a topic of food safety (Wild and Hall, 2000).

## 12.5 GMO Crops

Chemical control of pests is widely practiced, but more and more the use of crops and animals genetically improved to resist pests (insects, disease, nematodes, and weeds) are being explored and developed to offset chemical usage while protecting valuable food sources, in wheat, rice, and with many other crop staples and in food animals. In some cases, resistance genes are engineered into the crop, giving farmers new genetic resources for insect resistance (e.g., Bt toxin genes in corn and soybeans). In these improved varieties, little or no external chemical insecticide application may be needed. Other crops (papaya and plums) have been made resistant to viral diseases by transgenically imparting production of viral coat proteins that stop virus reproduction (Lindbo and Falk, 2017).

Gene-based technologies, such as RNA interference (RNAi), are underpinning new technologies in pest control (Kupferschmidt, 2021; Zhu, 2013). RNAi is a natural process that affects the activity of genes. Research has successfully led to artificial RNAs that target genes in pest insects, slowing growth or killing them. The development of GMO crops that make RNAi harmful to their pests is under active exploration. As with most new technologies, there are safety concerns that RNAi might also harm desirable species.

Jennifer Doudna and Emmanuelle Charpentier were awarded the 2020 Nobel Prize in Chemistry for their development of the CRISPR/Cas9 genome editing technology (Cross, 2020). Tools have already been developed in crops, livestock, and medical sciences (sickle cell disease, muscular dystrophy, COVID-19 diagnostic tests) (Cross, 2020). Plant scientists are using CRISPR gene editing (Bomgardner, 2017) to make sustainable agriculture crops with higher precision than possible before and less potential for undesirable side effects. While these new technologies like gene editing will have a huge impact in agriculture and in design of new drugs, it is not clear whether they will be embraced by consumers, at least in the case of food products. Some environmental organizations have indicated that they will resist introduction of new crops and farm animals improved by gene modification. Even though new technologies for genetic modification of crops offer the promise of safer and more abundant foods, many consumers and activists are also interested in preserving the natural qualities of foods, including taste, texture, color, and growth-related characteristics that influence availability and market choice.

Seemingly, for every technological advance in developing resistance, the target pest evolves a strategy for overcoming the protection, as happened so often with resistance in insect and fungal pests previously controlled with synthetic pesticides and with antibiotic use in farm animals. This is possible with genetically modified crops and biopesticides, thus requiring close monitoring of fields for early signs of resistance, and then applying an alternative pest control strategy from a "tool box approach" which may include conventional chemical pesticides, biopesticides, cultural methods, and other approaches to preserve these desirable, new technologies.

## 12.6 Smart Application Systems

Only a small fraction of applied pesticides reaches the intended pests, but rather bypass the targets and enter the soil, nontarget vegetation, or are carried away by wind (Duke, 2017). Agricultural engineers and systems scientists have developed more effective spraying techniques using spray drift control technology, electrostatically charged spray droplets, controlled release technology, or smart systems that direct spray just to the optimal position for contacting the target. These improvements save on the amount of pesticide needed for a particular situation and also prevent inadvertent residues that can harm unintended crops, water way quality, or livestock and wild animals. But they can also lead to more effective control of the target pest. Related to this, information is more accessible that clearly delineates the mere presence of a pest from the presence of a pest population of sufficient magnitude that can cause economic damage or impair safety in the harvested commodity. Not spraying at all can sometimes be the best strategy, generating a cycle of pest control by beneficial natural enemies, a means of biocontrol that, once fostered, can lead to sustainable pest suppression for many years (Monosson, 2017; Schar, 2017). This can have important ramifications in safety of foods, as lower chemical use may lead to safer foods as well as foods that are more widely available at relatively low costs.

Recent discussions with farming groups and food producers in China's Weifang Food Valley area of Shandong Province illustrate local efforts to reduce the use of conventional pesticides and fertilizers, including

- Develop farming practices for organic produce and meat
- Encourage use of biopesticides and biofertilizers
- Control nematodes by nonchemical means, such as soil solarization
- Develop novel ways of reducing development of resistance to pesticides
- Introduce nonchemical insect and disease controls by biological or cultural means
- Explore the Traditional Chinese Medicine (TCM) approach to find new antibiotics for use in farm animals and pest control
- Assist in adoption and registration of biopesticides using an IR-4 type regional or national system such as in use in the United States (IR-4, 2021)

## 12.7 Conclusions

The field of pest management, in agriculture, public health, and home use, is undergoing change as sole reliance on synthetic chemicals gives way to a variety of approaches, some with little or no pesticide in the mix. Health concerns over exposures to pesticides among agricultural workers and consumers of residue-tainted food products, as well as wildlife and nontarget species, have been partly responsible for this, as have demands by consumers, translated through the food supply chain to large commercial retail outlets, commodity organizations, and regulatory agencies, for pesticide free, organic, and "green" products. Still, the reality is that feeding a world population expected to top 9 billion by 2050 will require the use of chemical pesticides as a primary tool in combatting pests in the field and in stored products, as well as for public health, for years to come. It may, in fact, increase use of pesticides as emerging economies of China, India, Brazil, and other nations expand agricultural production. But the options for pest control have expanded, also illustrated by biopesticides, and advances in the fields such as genetics, biotechnology, sustainability, and more targeted pesticide application technology. These factors may in the long run reduce the need for chemical pest management tools to a fraction of that used presently.

Other transformative research opportunities that will require mathematical and physical science research efforts to improve the sustainability of agriculture include (NSF Advisory Committee, 2014)

- ensuring a sustainable water supply for agriculture,
- closing the loop for nutrient life cycles,
- crop protection as noted in this manuscript,
- innovations to prevent waste of food and energy,
- sensors for food security and safety, and
- maximizing biomass conversion to fuels, chemicals, food, and biomaterials (see Chapter 13, Section 13.1, NEXUS APPROACH)

This chapter was reproduced and adapted with permission from Seiber, J.N., Coats, J., Duke, S.O., and Gross, A.D. (2018). Pest management with biopesticides. Front. Agri. Sci. Eng., Vol. 5, Issue 3, pp 295–300. © The Author(s) 2018. Published by Higher Education Press. This is an open access article under the CC BY license (http://creativecommons.org/licenses/by/4.0).

## References

Bomgardner. (2017). CRISPR: A new toolbox for better crops. *Chem. Eng. News.*

California Department of Pesticide Regulation. (2017). Pesticide use reporting, annual 2017. 95(24), 30–34. https://www.cdpr.ca.gov/docs/pur/pur17rep/17_pur.htm (Accessed 2/3/21).

Cantrell, C.L., Dayan, F.E., and Duke, S.O. (2012). Natural products as sources for new pesticides. *J. Nat. Prod.* 75, 1231–1242.

Chang, C.L., Cho, I.K., and Li, Q.X. 2009. Insecticidal activity of basil oil, trans-anethole, estragole, and linalool to adult fruit flies of *Ceratitis capitata*, *Bactrocera dorsalis*, and *Bactrocera cucurbitae*. *J. Economic Entomol.* 102(1): 203–209.

Cross, R. (2020). CRISPR genome editing gets 2020 nobel prize in chemistry. *Chem. Eng. News.* 98(39).

Dayan, F.E., and Duke, S.O. (2010). Natural products for weed management in organic farming in the USA. *Outlooks Pest Manag.* 21, 156–160.

Dayan, F.E., Duke, S.O., Sauldubois, A., Singh, N., McCurdy, C., and Cantrell, C. (2007). p-Hydroxyphenylpyruvate dioxygenase is a herbicidal target site for β-triketones from *Leptospermum scoparium*. *Phytochemistry* 68, 2004–2014.

Dayan, F.E., Owens, D.K., and Duke, S.O. (2012). Rationale for a natural products approach to herbicide discovery. *Pest Manag. Sci.* 68, 519–528.

Duke, S.O. (2012). Why have no new herbicide modes of action appeared in recent years? *Pest Manag. Sci.* 68, 505–512.

Duke, S.O. (2017). Pesticide dose–A parameter with many implications. In Keith R. Solomon, Per Kudsk, Stephen O. Duke (Eds.), *Pesticide Dose Effects on the Environment and Target and Non-Target Organisms*. (pp. 1–13). United States: ACS Publications.

Griffith, C.M., Woodrow, J.E., and Seiber, J.N. (2015). Environmental behavior and analysis of agricultural sulfur. *Pest Manag. Sci.* 71, 1486–1496.

Heap, I., and Duke, S.O. (2018). Overview of glyphosate-resistant weeds worldwide. *Pest Manag. Sci.* 74, 1040–1049.

IR-4 (2021) IR-4 Project. https://www.ir4project.org/ (Accessed 2/3/21).

Isman, M.B., and Grieneisen, M.L. (2014). Botanical insecticide research many publications, limited useful data. *Trends Plant Sci.* 19, 140–145.

Kupferschmidt, K. (2021). Viral evolution may herald new pandemic phase. *Science* 371, 108–109.

Leahy, J., Mendelsohn, M., Kough, J., Jones, R., and Berckes, N. (2014). Biopesticide oversight and registration at the US environmental protection agency. *Biopestic. State Art Future Oppor.* 1172, 3–18.

Li, A.S., Iijima, A., Huang, J., Li, Q.X., and Chen, Y. 2020. Putative mode of action of the monoterpenoids linalool, estragole, methyl eugenol and citronellal on ligand-gated ion channels. *Engineering* Doi: 10.1016/j.eng.2019.07.027.

Li, Q.X., and Chang, C.L. 2016. Basil (*Ocimum basilicum* L.) oils. In: V. Preedy (Ed.), *Essential Oils in Food Preservation, Flavor and Safety*. (pp. 231–238). London: Elsevier.

Light, D.M., and Beck, J.J. (2010). Characterization of microencapsulated pear ester, (2 E, 4 Z)-ethyl-2, 4-decadienoate, a kairomonal spray adjuvant against neonate codling moth larvae. *J. Agric. Food Chem.* 58, 7838–7845.

Lindbo, J.A., and Falk, B.W. (2017). The impact of "coat protein-mediated virus resistance" in applied plant pathology and basic research. *Phytopathology* 107, 624–634.

Monosson, E. (2017). *Natural Defense: Enlisting Bugs and Germs to Protect Our Food and Health*. Washington, D.C.: Island Press.

NSF Advisory Committee. (2014). *National Science Foundation-Mathematical and Physical Sciences Advisory Council Report: Crop Protection*. Washington, D.C.: National Science Foundation Advisory Committee Report.

Ritter, S.K. (2012). Pesticides trend all-natural. *Chem. Eng. News.* 90(36).

Sampson, B.J., Tabanca, N., Kirimer, N.E., Demirci, B., Baser, K.H.C., Khan, I.A., Spiers, J.M., and Wedge, D.E. 2005. Insecticidal activity of 23 essential oils and their major compounds against adult *Lipaphis pseudobrassicae* (Davis) (Aphididae: Homoptera). *Pest Manage. Sci.* 61, 1122–1128.

Schar, D. (2017). Unlikely allies. https://science.sciencemag.org/content/356/6343/1130/tab-figures-data.

Science (2013). Special issue smarter pest control. *Science* 341, 728–765.

Seiber, J.N., Coats, J., Duke, S.O., and Gross, A.D. (2018). Pest management with biopesticides. *Front. Agri. Sci. Eng.* 5(3): 295–300.

United Nations Environment Programme (2018). About montreal protocol. http://www.unenvironment.org/ozonaction/who-we-are/about-montreal-protocol (Accessed 2/3/21).

Wikipedia (2021). Mesotrione.

Wild, C.P., and Hall, A.J. (2000). Primary prevention of hepatocellular carcinoma in developing countries. *Mutat. Res. Mutat. Res.* 462, 381–393.

Williams, C.M. (1967). Third-generation pesticides. *Sci. Am.* 217, 13–17.

Zhu, K.Y. (2013). RNA interference a powerful tool in entomological research and a novel approach for insect pest management. *Insect Sci.* 20, 1–119.

# 13

## Conclusions

### 13.1 Conclusions

Pesticides are omni present in the environment and the major source is agricultural activities. Half or more of the chemicals released during pesticide application enter the air. Application and all other steps—plowing/tilling, irrigation, planting, harvesting, and agricultural residue combustion (a waning agricultural practice, banned in some places) —release vapors, dust, smoke, and particulate matter that contain pesticides and other by-products of combustion (e.g., PAHs, dioxins, and others). These forms of release, as well as those due to manufacturing of pesticides and waste disposal, are relatively predictable and manageable and thus give us a start on accounting for pesticides in the environment.

However, unexpected accidental spills and natural disasters often result in environmental and health catastrophes. For example, following the industrial accidental release of methyl isocyanate at Bhopal, India, killed an estimated 20,000 people and exposed another 200,000, and a railroad derailment near Mount Shasta dumped methyl isothiocyanate into the Sacramento River, killing trout, salmon, and other fish. Wildfires, another source of pesticides and combustion products in the environment, have become more prevalent due to global climate change. In 2020, the worst wildfires in U.S. history burned vast expanses of urban, agricultural, and forest land in California and along the west coast.

Multiple factors cause pesticides to reach unintended sources. The environmental mobility of pesticides, including their presence in air, is largely dictated by the physicochemical properties of the pesticide active ingredients and its formulation ingredients—vapor pressure (most important), solubility, adsorption, and reactivity. Initial deposits, or residues from application, can become mobilized by tillage, irrigation, and a myriad of human activities in the immediate area of use and remote areas (depending on climatic and topographical features). Mass transport (via wind and erosion) and atmospheric deposition (rain, fog, and dry deposition) also determine how much pesticide is in the air. The manmade problem of global climate change that increases the temperature and severity of weather events further magnifies

DOI: 10.1201/9781003217602-13

mobilization of pesticides. The frequency and severity of violent or extreme weather events, indicators of global warming, continue through 2020. For example, 29 storms were named in the Atlantic Ocean in 2020, breaking the previous record set in 2005. Unusually warm Atlantic waters and a La Nina event in the Pacific Ocean contributed to the stormy season (Segarra, 2020). Spread, sprayed, or applied in some other way, pesticides and other contaminants release to air inevitably occurs, resulting in airborne exposures to applicators and others in the vicinity of the application.

The point is that pesticides and toxics get into the air. The effects of pesticides on people, domestic animals, wild animals (including microorganisms and fish), and ecosystems are the result of exposures—amounts and duration. Regulations, usage practices, care, and safety all play roles in minimizing risk (exposure×hazard). Underlying all of this is the continuing need for research, training, and communication.

Taken one by one, these mobilization processes can be studied, understood, and modelled. But in the real world, all of these processes can occur simultaneously making it difficult to know how much exposure is occurring, and thus the magnitude of harmful effects. It is this dynamic that leads to approximations, worst case estimates, and attempts to limit or control exposures to airborne toxics, pesticides, and other contaminants that justify attention.

At the time of this writing, the world is riveted by a pandemic caused by the novel coronavirus, COVID-19. The same factors leading to human exposures to pesticides (physicochemical properties, environmental conditions, and human activities) exist for pathogenic viruses, fungi, and bacteria, as well as other chemicals, manufactured and naturally occurring—fuels, industrial chemicals, solvents, combustion products, freons and other fluorocarbons, toxic wastes, PCBs, solvents, and drugs, alkaloids, and detergents. Thus, much of what we have learned from pesticides helps in understanding and mitigating effects of COVID-19 and other air contaminants. The tools of the environmental toxicologists' trade—analytical methods, models, effects evaluation in humans and animals, forensics, epidemiology, and ecology— pertain broadly to addressing these other contaminants in air.

The world has experienced steady exponential growth in its population since 1940, the year of author JNS's birth (Figure 13.1). The population of earth in 2020 was 7,800,000,000 people. The total land mass (including areas covered by ice) is about 149,000,000 km$^2$. On average, this means that each person has 0.019 km$^2$ to acquire the needed resources for their life. In 1940, it was 0.0645 km$^2$/person. The amount of area per person is shrinking as the population rises. Will population growth continue on the same exponential path, will it level off to some steady state, or will population decline? These questions are yet to be answered and depend on how we address sustainability in the future—at our current pace of population growth and our current food, energy, and water practices, the planet will not have sufficient resources to meet these demands much longer.

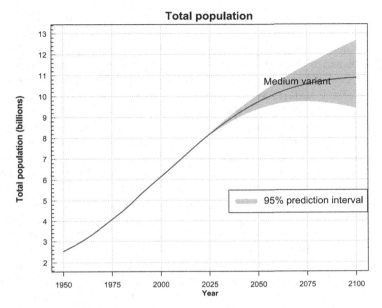

**FIGURE 13.1**
Projected increase in the world population (United Nations, 2020) (Credit: Copyright 2019 by United Nations, made available under a Creative Commons license (CC BY 3.0 IGO) http://creativecommons.org/licenses/by/3.0/igo/.)

We should plan now and move forward with major improvements to areas ranging from pest control to food production, food safety and healthfulness, and mitigation and reversal of climate change. It's all about "sustainability"— an often used but misunderstood term. Sustainability has different meanings, depending on the context. And sustainability from different perspectives is often antithesis.

Thus, a new interdependent way of thinking must be adopted as we move forward. In the food, energy, and water (FEW) Nexus, each of the elements essential to our survival—FEW—are interdependent. Much has been written and discussed regarding the nexus (Finley and Seiber, 2014; Urbinatti et al., 2020) and the biobased economy. Thinking within the FEW nexus paradigm, development of new technology in any sectors must consider the impact on the other sectors as well as the environment. Specifically, the development of a new energy or food crop must consider water use, energy consumption, and waste or by-product production during the process (Finley and Seiber, 2014).

Developing environmentally stable energy processes and sources can have a positive impact on the FEW nexus paradigm. Much progress has been made in the utilization of solar, wind, and nuclear sources of energy. More efficient carbon economy and air quality practices will benefit from improved energy storage technology and uses of biofuels. Other areas to improve the FEW nexus include reducing HCFC refrigerants through more environmentally favorable substitutes (see Chapter 9). Current agricultural practices rely on unstainable usage of water. Biotechnology offers one of the means of rapidly improving the drought tolerance of crops and opening new growing environments to meet the challenges of climate change. Approaches to the FEW Nexus can be seen in green chemistry and green crop production (Song et al., 2020). We are moving away from the era of the synthetic revolution into an era of green crop protection, which relies on a healthy, functioning ecosystem, resilient crops, and biopesticides. Furthermore, recent advances in greener technologies have created a dynamic FEW nexus landscape. The balance of water requirements and greenhouse gases produced by animals is an example of FEW nexus dynamics that are impacted by population growth and food preferences.

Opportunities abound in interdisciplinary education, international cooperation, and economic opportunities. At the center of this is the need to attract and educate our best and brightest to the opportunities that await. The book, *Leadership in Agriculture: Case Studies for a New Generation*, is a high-level view of how research in agriculture is done in the United States (Jordan et al., 2013). This system is coordinated at the highest levels. It includes the extension services in each state, which serves as outreach to industry, and other federal and state agencies like California Department of Pesticide Regulation and the U.S. Environmental Protection Agency. We hope that the topics on air quality and human and ecological health in this book will inspire broader improvements that can attract the coming generation to environmental improvement and—yes—sustainability.

Our best safeguard of sustainability is expanding knowledge. "The more we know, the more we grow!" that old saying is true now more than ever. The bottom line is we need research, education, and public policies that will continue to keep us focused on sustainability. And we need an educated populace able and willing to engage! It is in this sense that we offer this book and invite your feedback and suggestions.

Stay involved! Now more than ever.

## References

Finley, J.W., and Seiber, J.N. (2014). The nexus of food, energy, and water. *J. Agric. Food Chem.* 62, 6255–6262.

Jordan, J.P., Buchanan, G.A., Clarke, N.P., and Jordan, K.C. (2013). Leadership in agriculture: Case studies for a new generation (College Station: Texas A&M University Press).

Segarra, C. (2020). With Theta, 2020 sets the record for most named Atlantic storms. *Science News.* https://www.sciencenews.org/article/2020-storm-theta-record-atlantic-hurricane-climate (Accessed February 4, 2021).

Song, B., Seiber, J.N., Duke, S.O., and Li, Q.X. (2020). Green plant protection innovation: challenges and perspectives. *Engineering* 6, 483–484.

United Nations. (2020). World population prospects 2019- Volume II: demographic profiles. https://doi.org/10.18356/7707d011-en

Urbinatti, A., Benites Lázaro, L.L., Carvalho, C., and Giatti, L. (2020). The conceptual basis of water-energy-food nexus governance: systematic literature review using network and discourse analysis. *J. Integr. Environ. Sci.* 17, 1–23.

# Index

Printed in the United States
by Baker & Taylor Publisher Services